Symmetry: A Very Short Introduction

VERY SHORT INTRODUCTIONS are for anyone wanting a stimulating and accessible way into a new subject. They are written by experts, and have been translated into more than 45 different languages.

The series began in 1995, and now covers a wide variety of topics in every discipline. The VSI library now contains over 500 volumes—a Very Short Introduction to everything from Psychology and Philosophy of Science to American History and Relativity—and continues to grow in every subject area.

Titles in the series include the following:

AFRICAN HISTORY John Parker and
    Richard Rathbone
AGEING Nancy A. Pachana
AGNOSTICISM Robin Le Poidevin
AGRICULTURE Paul Brassley and
    Richard Soffe
ALEXANDER THE GREAT
    Hugh Bowden
ALGEBRA Peter M. Higgins
AMERICAN HISTORY Paul S. Boyer
AMERICAN IMMIGRATION
    David A. Gerber
AMERICAN LEGAL HISTORY
    G. Edward White
AMERICAN POLITICAL
    HISTORY Donald Critchlow
AMERICAN POLITICAL PARTIES
    AND ELECTIONS L. Sandy Maisel
AMERICAN POLITICS
    Richard M. Valelly
THE AMERICAN PRESIDENCY
    Charles O. Jones
AMERICAN SLAVERY
    Heather Andrea Williams
THE AMERICAN WEST Stephen Aron
AMERICAN WOMEN'S HISTORY
    Susan Ware
ANAESTHESIA Aidan O'Donnell
ANARCHISM Colin Ward
ANCIENT EGYPT Ian Shaw
ANCIENT GREECE Paul Cartledge
THE ANCIENT NEAR EAST
    Amanda H. Podany
ANCIENT PHILOSOPHY Julia Annas

ANCIENT WARFARE Harry Sidebottom
ANGLICANISM Mark Chapman
THE ANGLO-SAXON AGE John Blair
ANIMAL BEHAVIOUR
    Tristram D. Wyatt
ANIMAL RIGHTS David DeGrazia
ANXIETY Daniel Freeman and
    Jason Freeman
ARCHAEOLOGY Paul Bahn
ARISTOTLE Jonathan Barnes
ART HISTORY Dana Arnold
ART THEORY Cynthia Freeland
ASTROPHYSICS James Binney
ATHEISM Julian Baggini
THE ATMOSPHERE Paul I. Palmer
AUGUSTINE Henry Chadwick
THE AZTECS Davíd Carrasco
BABYLONIA Trevor Bryce
BACTERIA Sebastian G. B. Amyes
BANKING John Goddard and
    John O. S. Wilson
BARTHES Jonathan Culler
BEAUTY Roger Scruton
THE BIBLE John Riches
BLACK HOLES Katherine Blundell
BLOOD Chris Cooper
THE BODY Chris Shilling
THE BOOK OF MORMON
    Terryl Givens
BORDERS Alexander C. Diener and
    Joshua Hagen
THE BRAIN Michael O'Shea
THE BRICS Andrew F. Cooper
BRITISH POLITICS Anthony Wright

Ian Stewart

SYMMETRY

A Very Short Introduction

OXFORD
UNIVERSITY PRESS

OXFORD
UNIVERSITY PRESS

Great Clarendon Street, Oxford, ox2 6dp,
United Kingdom

Oxford University Press is a department of the University of Oxford.
It furthers the University's objective of excellence in research, scholarship,
and education by publishing worldwide. Oxford is a registered trade mark of
Oxford University Press in the UK and in certain other countries

© Joat Enterprises 2013

The moral rights of the author have been asserted

First Edition published in 2013

British Library Cataloguing in Publication Data

Data available

ISBN 978-0-19-965198-6

Printed and bound by
CPI Group (UK) Ltd, Croydon, CR0 4YY

Contents

List of illustrations

Introduction

Symmetry is an immensely important concept. A fascination with symmetric forms seems to be an innate feature of human perception, and for millennia it has influenced art and natural philosophy. More recently, symmetry has become indispensable in mathematics and science, where its applications range from atomic physics to zoology. Einstein's principle that the laws of Nature should be the same at all locations and all times, which forms the basis for fundamental physics, requires those laws to possess corresponding symmetries. But for thousands of years, the concept of symmetry was just an informal description of regularities of shape and structure. The main example was bilateral or mirror-image symmetry—for example, human bodies and faces look almost the same as their reflections. Occasionally the term was also used in connection with rotational symmetry, such as the fivefold symmetry of a starfish or the sixfold symmetry of a snowflake. The main focus was on symmetry as a geometric property of shapes, but sometimes the word was invoked in a metaphorical sense: for example, that in social disputes, both sides should be treated in the same way. The deeper implications of symmetry could not be discovered until the concept was made precise. Then mathematicians and scientists would have a solid base from which to investigate how symmetry affects the world we live in.

Today's formal concept of symmetry did not come from art or sociology. It did not come from geometry, either. Its primary source was algebra, and it emerged from a study of the solution of algebraic equations. An algebraic formula has symmetry if some of its variables can be interchanged without altering its value. In the 1800s several mathematicians, notably Niels Henrik Abel and Évariste Galois, were attempting to understand the general equation of the fifth degree. They proved, in two related but different ways, that this equation cannot be solved by any formula of the traditional kind ('radicals'). Both analysed the relation between such a solution and symmetric functions of the roots of the equation. What emerged was a new algebraic concept: a group of permutations.

After an initial hiatus while mathematicians got used to this new idea, it soon became apparent that structures remarkably similar to groups of permutations occurred naturally in many different areas of mathematics, not just algebra. Among these areas were complex function theory and knot theory. General and more abstract definitions of a group appeared, and a new subject was born: group theory. At first most work in this area was algebraic, but Felix Klein pointed out a deep connection between the concepts that made sense in any specific type of geometry and the group of transformations upon which that geometry was based. This connection allowed theorems to be transferred from one area of geometry to another, and unified what at the time was an increasingly disparate collection of geometries—Euclidean, spherical, projective, elliptic, hyperbolic, affine, inversive, and topological.

At much the same time, crystallographers realized that group theory could be used to classify the different types of crystal, by considering the symmetries of the crystal's atomic lattice. Chemists began to understand how the symmetries of molecules affected their physical behaviour. General theorems linked symmetries of mechanical systems to the great classical conserved quantities, such as energy and angular momentum.

Symmetry is a highly visual topic with many applications, such as animal markings, locomotion, waves, the shape of the Earth, and the form of galaxies. It is fundamental to both of the core theories of physics, relativity and quantum theory, and provides a starting point for the ongoing search for a unified theory that subsumes them both. This makes the topic ideal for a Very Short Introduction. My aim is to discuss the historical origins of symmetry, some of its key mathematical features, its relevance to patterns in the natural world, including living organisms, and its applications to pattern formation and fundamental physics.

The story begins with simple examples of symmetry related to everyday life. These lead to the great breakthrough: the realization that objects do not have symmetry: they have symmet*ries*. These are transformations that leave the object unchanged. This concept then extends to symmetries of more abstract entities, such as mathematical equations and algebraic structures, leading to the general notion of a group. Some of the basic theorems of the subject are then stated and motivated, without proofs.

Next, we describe some of the many different types of symmetry—translations, rotations, reflections, permutations, and so on. In combination, these transformations lead to many symmetric structures that are vital in both mathematics and science: cyclic and dihedral symmetry, frieze patterns, lattices, wallpaper patterns, regular solids, and crystallographic groups. For light relief, we discuss how group theory can be applied to some familiar games and puzzles: the Fifteen Puzzle, the Rubik cube, and sudoku.

Equipped with a refined understanding of symmetry, we examine how Nature's patterns, especially familiar ones from everyday life,

3

can be described and explained through symmetry. Examples include crystals, water waves, sand dunes, the shape of the Earth, spiral galaxies, animal markings, seashells, animal movement, and the spiral *Nautilus* shell. These examples motivate the concept of symmetry breaking, which is a general pattern-forming mechanism.

Delving deeper, we examine the profound impact that symmetry has had on the basic equations of mathematical physics. Symmetries of mechanical equations, now conceptualized as Lie groups, are closely related to fundamental conservation laws via Noether's Theorem. An important class, the simple Lie groups, can be classified completely. Lie groups appear in relativity and quantum mechanics, providing an entry route for the search for unified field theories—so-called Theories of Everything—such as string theory.

The groups that are vital to mathematical physics feed back into the mathematical foundations of symmetry in a surprising manner. Their study forms a key part of one of the great triumphs of 20th-century mathematics: the awe-inspiring classification of all finite simple groups. These turn out to be the alternating groups; finite analogues of simple Lie groups, in which the real or complex numbers are replaced by finite fields, plus some cunning variations; and twenty-six puzzling 'sporadic' groups, culminating in a truly remarkable and utterly enormous group called the monster.

Chapter 1
What is symmetry?

Three bored children on a ferry are passing the time by playing a game. It is a traditional game, requiring no apparatus beyond the children themselves: rock–paper–scissors. Make the shape behind your back with your hands, then reveal it. Rock blunts scissors. Scissors cut paper. Paper wraps rock.

In the distance, waves roll up a sandy beach, breaking as they reach the shore: an apparently endless succession of parallel ridges of water.

Half the sky is a layer of thick grey cloud as a summer shower falls. Illuminated by the bright sun in the other half of the sky, a polychrome rainbow arches across the heavens.

A schoolboy passes by on his bicycle, moving smoothly along the road.

He stops to watch the ferry docking. He feels guilty because he should be doing his geometry homework on isosceles triangles. Like generations of schoolboys before him, he is stuck at the *pons asinorum*—the bridge of asses. *Why* are the base angles equal? To him, Euclid's proof is opaque and inscrutable.

* * *

I put Euclid in as a broad hint that these scenes from everyday life have some kind of mathematical content. In fact, all five verbal snapshots have a common theme: symmetry. The children's game is symmetric: neither child has an advantage or a disadvantage, whichever choice they make. The waves rolling up the beach are symmetric: they all look pretty much alike. The rainbow is beautiful and elegantly proportioned, attributes often associated with symmetry in a metaphorical sense, but it has a more literal symmetry too. Its coloured arcs are circular, and circles are very symmetric indeed—which may be why ancient Greek philosophers considered circles to be the perfect form. Each wheel of the bicycle is also a circle, and it is the circle's symmetry that makes the bicycle work: perfection of form is subjective and irrelevant to mechanics, but symmetry is crucial. The schoolboy, trying to understand the mindset of an ancient Greek mathematician, is frustrated because he has not yet become aware of a hidden symmetry in Euclid's proof—one that would have reduced the whole problem to a single, obvious statement, had Euclid's culture allowed him to think that way.

I've used the word 'symmetry' many times already, but I haven't explained what it is—and now is too early. It's a simple yet subtle concept. A general definition will emerge from these examples, but, for now, let's consider each in turn, starting with the simplest and most direct.

Bicycle

Why are wheels circular? Because circles can roll smoothly. When a wheel rolls over a flat surface, successive positions look like Figure 1. The wheel rotates through an angle between each position and the next, but looking at the picture, you can't tell the difference. You can see that the circle has moved, but you can't see any difference in the circle itself. However, if you put a mark on the circle, you will see that it has rotated, through an angle that is

1. Why wheels work

proportional to the distance travelled. The wheel has circular symmetry: every point on the rim is the same distance from the centre. So it can roll along the flat surface, and the centre always stays at the same height. Just the place to put an axle.

Circles work on bumpy surfaces too, as long as the bumps are gentle or small enough not to matter. If you are given the luxury of redesigning the road, circular symmetry is neither necessary nor sufficient for a shape to roll. Square wheels work pretty well when the road is a series of upside-down catenaries, as in Figure 2 (left), although the motion is a bit jerky. In fact, given any shape of wheel, there exists a road that it can run on while staying level: see Leon Hall and Stan Wagon, 'Roads and wheels', *Mathematics Magazine* **65** (1992) 283–301. Non-circular shapes with constant width make poor wheels but perfectly good rollers. The simplest is constructed by swinging arcs of circles from the corners of an equilateral triangle, shown in white in Figure 2 (right).

2. *Left*: A road fit for square wheels. *Right*: Any curve of constant width can be used as a roller

Rainbow

Why do rainbows look the way they do? Everyone focuses on the colours, and we've all been told the answer: a drop of water is like a prism, and prisms split white light into its constituent colours. But what about the shape? Why is a rainbow formed from a series of bright bands, forming a great arch in the sky? Ignoring the shape of the rainbow is like explaining why a fern is green but not why it's fern shaped.

The main problem with the usual explanation of the rainbow is that although each droplet of water acts like a prism, despite not being shaped like one, a rainbow involves millions of droplets spread over a large volume of space. Why don't all those coloured rays get in each other's way, producing a muddy smeared-out pattern? Why do we see a concentrated band of light? Why do the different colours stand out?

The answer lies in the geometry of light passing through a spherical droplet. (Incidentally, you also need to understand the geometry to see why a droplet works like a prism, since it has no sharp corners.) Imagine a tight bunch of parallel light rays from the sun, encountering a single tiny droplet. Each ray is really a combination of rays of many distinct colours—as the prism experiment shows—so it simplifies the problem if at first we consider just one colour. The incoming light bounces round inside the droplet and is reflected back. What happens is surprisingly complex, but the main feature, which creates the rainbow, is described by rays that hit the front of the droplet, pass inside and are refracted by the water, hit the back of the droplet and are reflected, and finally pass out again through the front, being further refracted. This is not as straightforward as passing in through one face of a prism and out through the opposite one.

The geometry of this process is illustrated in Figure 3 for incoming rays lying in a plane through the line that joins the

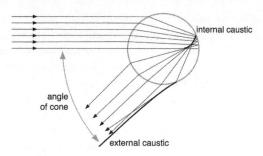

internal caustic

angle
of cone

external caustic

3. Geometry of the rainbow

centre of the sun to the centre of the droplet. This line is an axis of
rotational symmetry for the entire system of rays. The main
features are two *caustics*: curves to which the rays are all tangent.
Caustics are the places where the light is concentrated, a kind of
focusing effect. The name means 'burning', which is what sunlight
passing through a lens will do to skin. One caustic lies inside the
droplet, and the other is outside. The external caustic is
asymptotic to a straight line at a specific angle to the axis of
symmetry. So for each colour, most of the light emitted by the
droplet is focused at a specific angle to the axis. Because the
system of rays is rotationally symmetric, the emergent rays lie very
close to a bright cone.

When we look at a rainbow, most of the light that we see comes
from those droplets whose cones happen to meet our eye. Simple
geometry shows that these droplets lie on another cone, with our
eye at the tip, pointing in exactly the opposite direction to the
cones emitted by the raindrops. It has the same vertex angle as the
cone of emitted light, and its axis is the line joining the sun to our
eye. So we observe a cross section of a cone, which is a bright
circular arc. The other raindrops don't smear that out because
hardly any of their light hits our eye.

What about the coloured bands? They arise because the angle of
refraction depends on the wavelength of the light. Different

wavelengths, corresponding to different colours, produce arcs of slightly different sizes. For visible light, the angle lies roughly between 40° (blue) and 42° (red). These arcs all have the same centre, which lies on the symmetry axis. There's much more to rainbows, for example the common occurrence of a secondary rainbow, which lies outside the main one, is not as bright, and has the colours in the reverse order. This is created by rays that bounce more than once inside the droplets. But the overall shape is a consequence of rotational symmetry, both of a droplet and of the entire system. Next time you see a rainbow, don't think prisms, think symmetry.

Ocean waves

In reality, waves rolling up the beach are not precisely identical, but in some circumstances they come close: for example, gentle ripples on a very calm sea. Simple mathematical equations for waves reproduce this pattern: they have regular periodic solutions. In the simplest model of all, with space reduced to one dimension and assuming the wave height to be small, a wave is a sine curve moving at constant speed; see Figure 4.

Sine curves have an important symmetry indicated by the arrows in the figure: they are periodic. Add 2π to any angle, and its sine remains the same. That is,

$$\sin(x + 2\pi) = \sin x$$

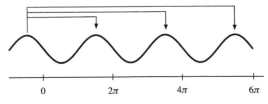

0	2π	4π	6π

4. Sinusoidal waves

So at any instant of time, the spatial pattern of the wave would look exactly the same if you slid the whole wave along by a distance 2π, or any integer multiple. And 'look exactly the same' is one of the characteristic features of symmetry.

Moving waves have another type of symmetry: symmetry in time. If the wave is travelling with speed c then its shape at time t is $\sin(x-ct)$. After time $2\pi/c$ that becomes $\sin(x-2\pi)$, which equals $\sin x$. So the pattern looks the same after a time that is any integer multiple of $2\pi/c$. This is why each successive wave looks much the same as the previous one.

In fact, a sine wave has even more symmetry: it maintains the same shape as it moves. If you slide the wave sideways by any amount a and wait for a time a/c, you see exactly the shape that you started with, because $\sin(x + a-ca/c) = \sin x$. This type of spatio-temporal symmetry is characteristic of travelling waves.

Rock–paper–scissors

In the previous examples, symmetry is associated with geometry. However, symmetry need not be related to anything visual. The symmetry of rock–paper–scissors is crystal clear, and everybody sees it immediately because it's what makes the game fair. All three strategies are 'on the same footing'. Whatever one child chooses, the other has one choice that beats it, one that loses to it, and one that is the same and therefore leads to a draw.

Rock–paper–scissors is a game in a more formal sense. In 1927 John von Neumann, one of the great mathematicians of the 20th century and a pioneer of computer science, invented a simple model of economic decision-making, called *game theory*. He proved a key theorem about games in 1928, and this led to an explosion of new results, culminating in *Theory of Games and Economic Behaviour*, written jointly with Oskar Morgenstern and published in 1944. It became a media sensation.

In the simplest version of von Neumann's set-up, a game is played by two people. Each has a specific set of available strategies, and must choose one of them. Neither player knows what their opponent is going to choose, but they both know how their gains and losses—*payoffs*—depend on the combination of choices that they make. In an economic application, one player might be a manufacturer and the other a potential customer. The manufacturer can choose what to make and what price to charge; the customer can decide whether or not to buy.

To bring out the basic mathematical principles, imagine that the two players repeat the same game many times, making new strategic choices on each repetition—just like the children on the ferry. Which strategy produces the greatest gain, or the least loss, on average? Always making the same choice is clearly a bad idea. If one child always chooses scissors, then the other can win every time by spotting the pattern and choosing rock. So von Neumann was led to consider *mixed strategies*, involving a range of random choices, each with a fixed probability. For example, choose scissors half the time, paper one-third of the time, and rock one-sixth of the time, at random. His basic result was the Minimax Theorem: for any game there exists a mixed strategy that permits both players simultaneously to make their maximum losses as small as possible. This result had been conjectured for some time, but it needed a proper proof, and von Neumann was the first person to find one. He said: 'There could be no theory of games...without that theorem...I thought there was nothing worth publishing until the Minimax Theorem was proved.'

The mixed strategy above is not minimax. If one player chooses scissors half the time, then the other can improve their chance of winning by choosing rock more frequently than paper. We can find the minimax strategy by exploiting the game's symmetry. Roughly speaking, the minimax strategy must have the same kind of symmetry. We can all guess where that leads to, but it will be useful to run through some of the details that confirm that guess.

Consider a mixed strategy in which a player chooses rock with probability r, paper with probability p, and scissors with probability s. Denote this strategy by (r, p, s) and suppose it is minimax. I'm going to use the symmetry of the game to deduce the values of r, p, and s.

First, we need a table of payoffs, called the *payoff matrix*. Scoring 1 for a win, –1 for a loss, and 0 for a draw, it looks like Figure 5 (left). I claim that if (r, p, s) is a minimax strategy for player 1, then so is (p, s, r). In fact, so is (s, r, p), but we don't use that. To see why, imagine renaming the choices according to the language of the aliens of Apellobetnees III, using the standard dictionary:

English	Apellobetnish
rock	payppr
paper	syzzrs
scissors	roq

The rules of the game sound the same in both languages—on Apellobetnees III, payppr beats roq beats syzzrs beats payppr. The payoff matrix looks the same whichever language we use. So the effect of this linguistic change is to cycle the strategies as in

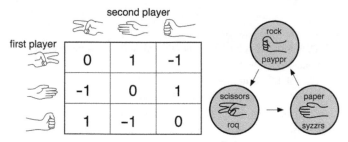

5. *Left*: **Payoff matrix for the first player in rock–paper–scissors.** *Right*: **Cycling strategies; arrow means 'beats'**

Figure 5 (right). The average gains or losses for any strategy (r, p, s) also don't change if we cycle the symbols, which leads to the strategy (p, s, r). Since these two strategies always have the same average gains and losses, it is clear that if one of them is minimax, so is the other.

Usually there is only one minimax strategy. I don't want to get tied up in the technicalities, but it's true for rock–paper–scissors. So the two mixed strategies are the same:

$$(r, p, s) = (p, s, r)$$

That means that $r = p = s$. But a player *must* choose one of the three shapes, so the probabilities sum to 1:

$$r + p + s = 1$$

Therefore r, p, and s all equal 1/3. In short: the minimax strategy for rock–paper–scissors is to choose each shape at random with equal probability.

As I said, you could have guessed this. But we now know why it's true—and which technical theorems you need to prove to demonstrate that. The mathematical skeleton of the argument ignores many details of the problem; instead, it focuses on general principles:

1. The problem is symmetric.
2. Therefore any solution implies the existence of symmetrically related ones.
3. The solution is unique.
4. Therefore the symmetrically related solutions are all the same.
5. Therefore the solution we require is itself symmetric, and that determines the probabilities.

14

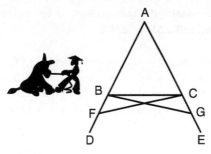

6. *Pons asinorum*

Bridge of asses

Euclid's proof that the angles at the base of an isosceles triangle
are equal is quite complicated. Likely reasons for its nickname are
the diagram, which resembles a bridge (see Figure 6), and its
metaphorical status as a bridge to the deeper theorems to which it
leads. Another, more frivolous, suggestion is that many students
ground to a halt when required to cross it.

Here's how Euclid proves the theorem. I've taken some liberties
and used *simpler* language, shortening the argument considerably.
I've abbreviated 'equal', 'angle', and 'triangle' in the usual way
($=$, $\angle$, Δ). Equality for triangles is what we now call 'congruent'—
same shape and size.

Let ABC be an isosceles triangle with AB = AC.

Extend AB and AC to get BD and CE. We claim that
$\angle$ABC = $\angle$ACB.

To prove this, take F somewhere on BD. From AE cut off AG = AF.
Draw FC and GB. Now FA = GA and AC = AB. ΔAFC and ΔAGB
contain a common angle $\angle$FAG. Therefore ΔAFC = ΔAGB, so

FC = GB. The remaining angles are equal in corresponding pairs: ∠ACF = ∠ABG and ∠AFC = ∠AGB.

Since AF = AG and AB = AC, the remainder BF = CG. Consider △BFC and △CGB. Now FC = GB and ∠BFC = ∠CGB, while the base BC is common to both triangles. Therefore △BFC = △CGB, so their corresponding angles are equal. Therefore ∠FBC = ∠GCB and ∠BCF = ∠CBG.

Since ∠ABG was proved equal to ∠ACF, and ∠CBG = ∠BCF, the remaining ∠ABC = ∠ACB, and these angles are at the base of △ABC.

QED

What's going on here? What is the *idea* behind Euclid's list of formal deductions?

The clue is that everything comes in matching left–right pairs. The sides AB and AC start this process; the angles we want to prove equal end it. F and G are symmetrically related; so are FC and GB; so are all the pairs of angles that are proved equal. Euclid compiles enough equal pairs of angles to conclude that ∠ABC = ∠ACB, as required. This, too, is a symmetrically related pair. Figure 7 illustrates the main steps, showing the symmetry throughout.

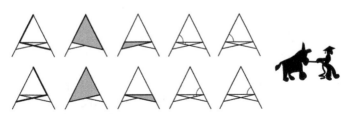

7. In each vertical pair of diagrams the marked lines, angles, and triangles are equal

16

Euclid's jumble of letters now starts to tell a mathematical story, the essence of a memorable, insightful proof. The essential idea, from a modern standpoint, is that the isosceles triangle has mirror symmetry. It looks the same if you reflect it in the vertical line through its apex. Since this operation swaps the base angles, they must be equal.

Why didn't Euclid prove it like that? He didn't have the luxury of referring to symmetry. The closest he could get was the notion of congruent triangles. He was building up his geometry step by logical step, and some concepts that seem obvious to us were not available at this stage of his book. So instead of flipping the triangle over to compare it to its mirror image, he constructed mirror-image pairs of lines and angles to do the same job, using the technical tool of congruent triangles to prove the required equalities.

Ironically, there is a very simple way to prove the theorem using congruent triangles, without adding any extra construction lines. Observe that $\triangle ABC$ is congruent to $\triangle ACB$. Two sets of corresponding sides are equal (AB = AC and AC = AB) and the included angles $\angle BAC$ and $\angle CAB$ are equal since they are *the same* angle.

Euclid didn't think like that, though. To him, $\triangle ABC$ and $\triangle ACB$ were the same triangle. What he needed was to define a triangle as an *ordered* triple of line segments. But he was thinking in pictures, and that level of abstraction was not available to him. I'm not saying he couldn't have done it. I'm saying that his cultural perspective didn't permit it.

* * *

We've now seen that symmetries of various kinds arise naturally in mathematics and the world around us, and that their presence can often simplify a calculation, provide insight into Nature, or motivate a proof. We've also seen that, mathematically, symmetry can be a shape (circles, waves), an abstract structure (rock–paper–

scissors), or a reflection (bridge of asses). The physical implications of symmetry can apply to space, to time, to both in combination, or to more abstract notions such as probability or a matrix.

What I've not yet explained is what symmetry *is*. The diversity of contexts in which the word seems applicable suggests that a precise definition could be elusive. However, the examples mainly indicate what symmetry is *not*. It isn't a number, it isn't a shape, it isn't an equation. It isn't space and it isn't time. It might be one of those metaphorical or judgemental ideas that you can't pin down formally, like 'beauty'. However, it turns out that there is a useful and precise notion of symmetry, broad enough to cover all of our previous examples and a great deal besides. Even more general notions of symmetry exist; this one isn't sacred. But it is extremely powerful and useful, and it's the industry standard in pure mathematics, applied mathematics, mathematical physics, chemistry, and many other branches of science.

When talking of the symmetry of a circle I described it in two ways. One was: every point is the same distance from the centre. The other was: if you rotate the circle through any angle, it looks exactly the same as it was to begin with. The second version is the one that holds the key to a formal definition of symmetry.

What is a rotation? Physically, it is a way to move an object by changing its orientation without changing its shape. Mathematically it is a transformation—an alternative word for 'function'. A transformation is a rule F that associates to any appropriate 'thing' x another 'thing' $F(x)$. The 'thing' might be a number, a shape, an algebraic structure, or a process. There's a fancy set-theoretic definition: if you know it I don't need to say what it is, and if you don't, you know enough already without it.

For the present example, x is a point on a circle, so I'll replace x by the more traditional symbol θ. Imagine the unit circle in the plane.

I can prescribe a point on the circle by letting θ be the angle at which it sits. What happens if I rotate the circle through, say, a right angle? Then the point θ moves to a new point at angle $\theta + \pi/2$. So this particular rotation can be defined using the transformation F for which:

$$F(\theta) = \theta + \pi/2$$

In these terms, what does my statement 'the circle looks exactly the same as it was to begin with' mean? Each *point* on the circle moves—it turns through a right angle. But the *set* of all rotated points is exactly the same as the original set—the circle. What's changed is how we label those points with angles.

More generally, rotation through a general angle α corresponds to ('is', in fact, by definition) the transformation

$$F_\alpha(\theta) = \theta + \alpha$$

which adds the same angle α to every angle θ representing a point on the circle. Again, the set of all rotated points is exactly the same as the original set. We say that a circle is symmetric under all rotations.

* * *

We can now define symmetry.

A symmetry of some mathematical structure is a transformation of that structure, of a specified kind, that leaves specified properties of the structure unchanged.

There is one technical condition: only invertible transformations, ones that can be inverted (reversed), are permitted. So we can't squash the entire circle down to a single point, for instance. Rotations are invertible: the inverse of rotation by α is rotation by $-\alpha$; that is, through the same angle, but in the opposite direction.

If the definition of symmetry seems a bit vague, that's because it's extremely general. 'Specified' is vague until you specify. For shapes in the plane or space, the most natural transformations to specify are rigid motions, which leave the distances between pairs of points unchanged. Other types of transformation are possible; for instance, topological ones, which can bend space, compress it, stretch it, but not break or tear it. But here we will confine attention to rigid motions, which allows a more explicit definition: *A symmetry of a shape in the plane (or space) is a rigid motion of the plane (or space) that maps the shape to itself.*

With these specifications, does a circle have any other symmetries? Yes: reflections. Any rigid motion of the plane that maps the circle to itself must map its centre to itself. Consider the unit circle in the plane, centred at the origin. By convention, angles are measured anticlockwise from angle 0, which is on the positive x-axis. If we reflect the plane in any straight line through the centre, a conceptual mirror, then the circle again maps to itself. For a horizontal mirror, the reflection is R_0, where

$$R_0(\theta) = -\theta$$

If the mirror is inclined at an angle α to the horizontal, the reflection is R_α, where

$$R_\alpha(\theta) = 2\alpha - \theta$$

With a bit more technique, we can prove that these rotations and reflections comprise all possible rigid-motion symmetries of the circle.

Notice that the circle has infinitely many symmetries: one infinite family of rotations, and a second infinite family of reflections. Other shapes may be less richly endowed. For example, an ellipse (with its axes in the usual position: one horizontal, the other vertical) has exactly four symmetries; see Figure 8 (left). These are: leave it alone, rotate by π, and reflect about the horizontal or vertical axes. In symbols, the transformations concerned are F_0, F_π, R_0, and $R_{\pi/2}$.

An equilateral triangle, centred at the origin and with one vertex on the horizontal axis as in Figure 8 (middle), has six symmetries: rotate through 0, $2\pi/3$, $4\pi/3$, or reflect in any of the lines bisecting the triangle. Symbolically, these are F_0, $F_{2\pi/3}$, $F_{4\pi/3}$, R_0, $R_{2\pi/3}$, and $R_{4\pi/3}$. Similarly a square, Figure 8 (right), has eight symmetries: F_0, $F_{\pi/2}$, F_π, $F_{3\pi/2}$, R_0, $R_{\pi/2}$, R_π, and $R_{3\pi/2}$.

As these examples illustrate, a given shape may have many different symmetries. So instead of considering individual symmetries, we need to think about them all. It turns out that the set of all symmetries of a given shape—or, more generally, some structure—has an elegant algebraic property. Namely, if we 'compose' two symmetries by performing the transformations in turn, the result is also a symmetry.

You can check this property for the above examples on a case-by-case basis, but there's an easier way. First, note that composing two rigid motions produces a rigid motion: if you leave the distance between two points unchanged, and then leave it

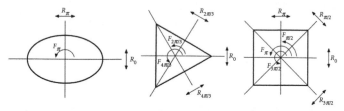

8. **Symmetries of an ellipse, equilateral triangle, and square.** F_0, which leaves all points fixed, is not indicated

unchanged again, you obviously leave it unchanged. Second, if each rigid motion concerned maps the shape to itself, then so does their composition: if you map a shape to itself and then map it to itself again, you have clearly mapped it to itself.

This property of symmetries is trivial, but it is also of vital importance. We say that the set of all symmetries of a given shape or structure forms a *group*. Accordingly, we rename this set the *symmetry group* of the shape or structure. It turns out that if you know the symmetry group, you can infer all sorts of things about the shape or structure. All five of my examples can be described in the language of symmetry groups, and the inferences that I made—circles make effective wheels, the minimax strategy in rock–paper–scissors is to choose each with the same probability, and so on—are applications of the appropriate symmetry group.

It is by no means evident that the symmetry group of a shape or structure is either useful or important, and the deductions I just mentioned can be carried out without any explicit references to symmetry. However, symmetry groups—and a more general concept, known simply as a group—turned out to be so useful that today little mathematics can be done without them. Historically, the group concept first appeared in a very important application, where no one had managed to make much progress without it, and once defined and understood, it blew the problem wide open: see Chapter 2. Only after that did mathematicians think about the general concept of symmetry and extract the definition of a symmetry group.

Chapter 2
Origins of symmetry

The broad notion of symmetry was tacitly recognized for
thousands of years before a specific version was formulated in
precise mathematical terms. Symmetries can be found in art,
culture, the natural world, science, and mathematics, and their
appeal seems to go to the roots of human perception. Religious
and secular symbols are often symmetric, and today the same is
true of some company logos; see Figure 9. A bold, simple,
symmetric design seems to have a powerful effect on the human
psyche. Artists have explored symmetric patterns in remarkable
depth and detail. Architects have employed a variety of
symmetries to design elegant buildings. Symmetry in the natural
world has fascinated natural historians and scientists since the
time of Aristotle.

Islamic art is famous for its use of symmetric designs, such as
those found in the Alhambra, a former fortress and palace in
Grenada, Spain that dates to the 14th century; see Figure 10.

9. Symmetric symbols and logos. *Left to right*: **Christianity, Judaism,
Islam, yin-yang, Mercedes, Toyota**

10. A typical Islamic pattern from the Alhambra

The building itself is not designed to a systematic plan, but it is decorated with a great many tiling patterns. There are seventeen distinct symmetry types of lattice pattern (Chapter 4), and it is often said that all of them exist in the Alhambra. The truth of this statement depends on how it is interpreted, because real patterns don't go on forever. Edith Muller found eleven (some say twelve) of them in 1944; Branko Grünbaum and others found two more in the 1980s but were unable to locate the remaining four. In 1987 Rafael Pérez-Gomes and independently José María Montesinos stated that they had succeeded in doing so. Grünbaum has disputed this claim on grounds of imprecise definitions. Syed Jan Abas and Amer Shaker Salman's *Symmetries of Islamic Geometrical Patterns* includes examples of all seventeen patterns from Islamic art, not all from the Alhambra. In addition, Islamic artists invented many patterns that at first sight appear to be perfectly symmetric, but cleverly avoid rigid mathematical obstacles to create 'impossible' patterns—for instance, containing apparently regular heptagons and octagons. Artistically, these are on a par with the perfectly symmetric patterns, and it seems unlikely that the artists distinguished between these two types

24

because they did not employ a rigorous mathematical characterization.

The most evident symmetry in Nature is the striking, though never exact, bilateral symmetry of animals, including humans. The spiral forms of many shells are another well-known instance of symmetry in the living world. Informal uses of symmetry, often tacit, occur throughout the sciences. For example, astronomers assumed that a large mass of molten rock in space was likely to assume a spherical form, and that a rotating mass would be axially symmetric.

More formal uses came in particular from crystallography. Crystals often have a striking geometric form; for example, salt crystals can be cubes. The facets of a crystal are evidence for the underlying atomic lattice, and the symmetries of the macroscopic crystal are related to those of this lattice. However, the details of the growth patterns of the crystal are also involved, in general, so the most direct predictions from the lattice are the angles between neighbouring facets. Historically, scientists only began to accept that crystals have a regular structure when measurements of these angles produced the same results for many samples of the same mineral. This may seem strange, but out in the field most mineral samples are damaged and fragmented, quite unlike the elegant specimens in museums. The symmetries of the atomic lattice are fundamental to the crystal's physical and chemical properties. Departures from symmetry are also important, but you have to know what they are departing *from*.

One of the pioneers of crystallography was Pierre Curie, and he stated his famous dissymmetry principle: an effect cannot have a dissymmetry absent from its cause. Dissymmetry is lack of symmetry, so we can rephrase this as 'The symmetry of the causes must be reproduced in the effects.' Suitably interpreted, this principle is arguably true, but its more obvious interpretations are often false; see Chapter 6. Crystallography was one of the first areas of science to benefit from a rigorous mathematical formulation of the concept of symmetry.

Another was chemistry, where it was discovered that many molecules exist in two mirror-image forms—the technical term is 'chirality', introduced in 1873 by Lord Kelvin. In 1815 Jean-Baptiste Biot noticed that some chemicals, notably sugars, rotate polarized light in one direction, while other apparently identical chemicals rotate it in the opposite direction. Louis Pasteur deduced in 1848 that the molecules concerned must be mirror images of each other. Chirality is important in biochemistry because one form of the molecule may be biologically active while the other is not. Amino acids, the basic constituents of proteins, are examples: the body can use one form but not the other. Many molecules are symmetric, and their properties are influenced by their symmetry. A recent case is buckminsterfullerene, a cage of 60 carbon atoms arranged like the vertices of a truncated icosahedron, which has the same symmetries as the icosahedron.

A rigorous definition of symmetry emerged from none of these areas. Instead, it came from pure mathematics, with the concept of a group of transformations. One of the basic rules in the history of mathematics seems to be that important, simple, general ideas first arise in a far more complicated form. Group theory is no exception. It emerged from a number of difficult technical areas of mathematical research, and in each case the concept had accompanying baggage: additional special structure that obscured the underlying simplicities. The historical sources for group theory include several distinct areas of algebra, of which the most influential is the theory of equations: how to solve, or in this case how not to solve, polynomial equations. Another source is the theory of elliptic functions and related 'modular' functions in complex analysis. An early application, knot theory, was also influential. Matrix algebra, which emerged from work on changes of variables in algebraic geometry, also played a major role, but we won't go into that.

You don't need to know any of this material to understand what a group is, or to use the concept. However, a feel for the history helps to put the material in context, and it demonstrates that what we are studying is real mathematics, related to core areas of the subject, and not just some weird abstraction with no purpose and no content.

Equations and Galois Theory

After basic geometry and arithmetic, the oldest area of mathematics is probably the theory of equations. Four thousand years ago, Babylonian scribes were teaching what in effect was today's elementary algebra to their pupils, using verbal descriptions and examples, often stated as 'I found a stone but did not weigh it…' followed by enough information to pin down the weight exactly. The students sat in classes, as they do today, and they were set homework. A few clay tablets even record their private impressions of their teachers, and those, too, are much as they would be today.

One triumph of Babylonian algebra, if I can call it that when no symbols were used, was the solution of quadratic equations. Babylon's scribes seem to have understood the general principle for solving any quadratic equation, although they presented their method through typical examples. The main difference today is the use of an algebraic formula to specify the solutions. Also, negative coefficients and complex solutions are now permitted. By the Renaissance, Italian mathematicians had discovered similar formulas for solving cubic and quartic equations. The common feature of these formulas was that, aside from the standard algebraic operations of addition, subtraction, multiplication, and division, the only extra ingredient was the extraction of nth roots—radicals. The solution of the quadratic required square roots; that of the cubic required both square roots and cube roots; that of the quartic also required square roots and cube roots. A fourth root, after all, is just the square root of a square root, so it's surplus to requirements.

The formulas were universal, in the sense that the same formula worked for any equation of the relevant degree. (In some circumstances the classical formula does not explicitly represent the real and imaginary parts of roots. This was first realized for cubic equations. If there is one real root, the formula provides it. If there are three real roots, it specifies them only as expressions involving cube roots of complex numbers.) However, as the degree increased, so did the complexity of the formula. For centuries there was a general feeling that the only obstacle to extending the results to equations of higher degree was this growing complexity. The solution of the general quintic equation, for instance, was presumably given by some complicated formula, no doubt involving fifth roots, cube roots, and square roots. It might possibly need seventh roots, or 107th roots for that matter, but it was hard to see why those would come into play.

By the late 18th century, some of the leading mathematicians were beginning to suspect that this belief was mistaken. Joseph Louis Lagrange found a unified description of previous methods for solving quadratics, cubics, and quartics; he used permutations of the roots of the equation to construct what we now call a Lagrange resolvent. This is a related equation, whose roots determine those of the original one. For quadratics, cubics, and quartics, the Lagrange resolvent has smaller degree than the original equation does. But for quintics, the Lagrange resolvent makes the problem worse, turning a quintic equation into a sextic equation—one of degree 6.

That does not imply that no solution by radicals exists. Perhaps there might be some alternative approach. Maybe Lagrange resolvents aren't the way to go. By 1799 the Italian mathematician Paolo Ruffini had written down what he claimed to be a proof that you do have to go that way, and it doesn't work. His title translates as 'The general theory of equations in which it is proved that the algebraic solution of equations of degree greater than 4 is impossible.' Unfortunately his book was enormously long, with extensive calculations where mistakes might easily occur, and the end result was negative, so his work attracted little attention. Ruffini tried to make his proof more accessible, but he never really got the credit he deserved. Later it turned out that his proof had a logical gap, but it could be filled.

The first accepted impossibility proof was published by the Norwegian Niels Henrik Abel in 1823, after an earlier episode where for a brief time he mistakenly thought he had found a formula that solved the quintic using radicals. His first proof was incomprehensibly brief, and in 1826 he published an expanded treatment. Like Lagrange and Ruffini, he focused on permutations of the roots of the equation. It was a proof by contradiction: assume there is a formula using radicals, and deduce something self-contradictory. The final step was a curious calculation involving two different permutations of the five roots.

The trouble with that sort of proof is that although you can check the logic and assure yourself that it is correct, it's not always clear *why* the answer is what it is. The big breakthrough came from the young Frenchman Évariste Galois, who made a head-on assault on the general question: when can a polynomial equation be solved by radicals? Galois gave a complete solution, which in passing proved that the general quintic cannot be solved in that manner, but it had what was seen at the time to be a flaw. It expressed the condition for solubility in terms of the roots, not the coefficients. This makes it difficult to verify Galois's conditions for any specific equation. Galois made things even worse for himself by becoming involved in revolutionary politics, and managed to get himself killed in a duel.

* * *

The techniques available to Galois were basic algebra and Lagrange's idea of a permutation. In those days, a permutation of a list of objects—say *abcde*—was another list in which they were rearranged, such as *bdaec*. This way of thinking, and writing, is rather cumbersome, but it was all Galois had. He associated with any polynomial equation a list of permutations of its roots, defined by certain algebraic properties, and showed that this list has a specific kind of structure. He called such a list a 'group'. And he proved that an equation can be solved by radicals if and only if its group can be broken up, in a particular manner, into a series of smaller groups, each having a very simple form. The idea was highly original, and it took time for the importance of the work to sink in.

In modern terms, his basic idea was to consider the *symmetry group* of the equation. Remember: a symmetry group consists of transformations that preserve certain structure. So what are the transformations, and what is the structure?

The transformations are permutations of the roots, but we now think of permutations as functions, not as arrangements. Instead of thinking of a standard list *abcde* and a rearrangement *bdaec*, we think of the transformation that replaces each symbol in the standard list by the corresponding one in the rearranged list. That is,

$$a \to b \quad b \to d \quad c \to a \quad d \to e \quad e \to c$$

This way of thinking has an advantage: it makes it obvious how to compose two permutations, and it is clear that this yields another permutation.

The structure that must be preserved is more subtle. It is not the *equation*. A permutation of the roots of an equation is just a reordering of those roots; the reordered roots satisfy exactly the same equation as they did in the original order. Instead, what must be preserved is *all algebraic relations* among the roots. Perhaps the original roots satisfy an equation like $ad-ce = 4$.

Apply the permutation, and this becomes $be-ac = 4$. If this relation does not hold, then the permutation is not a symmetry of the equation. If it does...well, there are other potential relations, and *all* of them have to be preserved. It's not immediately clear how to verify this condition, but it is clear that the permutations that satisfy it must form a group. We now call it the *Galois group* of the equation, and we define 'preserve relations' in a more abstract way.

This is where group theory, the mathematics of symmetry, came from. Not from geometric ideas about rotating squares or icosahedra. Everything would have been much clearer if the geometry had come first, and symmetry groups were available to Galois and his forerunners—but it didn't, and they weren't. Great pioneers never let this kind of thing stop them, but it does make their work harder for mere mortals to understand.

For a time, group theory was little more than an algebraic curiosity, important in just one area: the theory of equations. Undaunted, a few indefatigable pioneers continued developing group theory for its own sake. Soon groups littered the entire mathematical landscape. Henri Poincaré, in a slightly un-self-critical moment, once remarked that the theory of groups was, 'as it were, *the whole of mathematics* stripped of its matter and reduced to pure form' (my italics). The astonishing thing is not that he should make such a sweeping statement: it is that the statement was only a slight exaggeration. Groups had become that central and that important.

Areas where groups started making their presence felt included abstract algebra, topology, complex analysis, algebraic geometry, and differential equations. Connections with science, especially in physics and chemistry, also motivated further development of the group concept and its deep relationship with symmetry.

Abstract algebra

The modern abstract approach to algebra grew from the work of Galois and others on structural features of numbers, permutations, and similar systems. Galois himself investigated what we now call Galois fields: finite sets for which operations like 'addition' and 'multiplication' can be defined, satisfying all of the standard rules of algebra. There is one of these for each prime power number p^n of elements; it is denoted by $\mathbf{GF}(p^n)$.

The simplest example occurs when $n = 1$. Let $\mathbf{GF}(p)$ be the set of numbers $0, 1, 2, \ldots, p-1$, with operations:

$$a \oplus b = \text{the remainder on dividing } a+b \text{ by } p$$
$$a \otimes b = \text{the remainder on dividing } ab \text{ by } p$$

Then many familiar algebraic laws hold: for example, the commutative law for addition:

$$a \oplus b = b \oplus a$$

the distributive law:

$$a \otimes (b \oplus c) = (a \otimes b) \oplus (a \otimes c)$$

and simple rules like:

$$0 \oplus a = a \qquad 1 \otimes a = a$$

Moreover, when p is prime, every nonzero element a has a multiplicative inverse a^{-1} for which $aa^{-1} = 1$. So a^{-1} is effectively $1/a$ and we can define division:

$$a/b = ab^{-1}$$

For instance, suppose $p = 5$. Addition and multiplication tables look like this:

$\oplus$	0	1	2	3	4
0	0	1	2	3	4
1	1	2	3	4	0
2	2	3	4	0	1
3	3	4	0	1	2
4	4	0	1	2	3

$\otimes$	0	1	2	3	4
0	0	0	0	0	0
1	0	1	2	3	4
2	0	2	4	1	3
3	0	3	1	4	2
4	0	4	3	2	1

Now $2\otimes3 = 1$, so $2^{-1} = 3$ and $3^{-1} = 2$.

Number theorists had used this basic idea for some time before Carl Friedrich Gauss formalized it in his *Disquisitiones Arithmeticae*, introducing the notation:

$$x \equiv y \ (\mathrm{mod}\, n)$$

to mean that $x-y$ is exactly divisible by n. The resulting system is known as 'arithmetic modulo n'. When p is prime, so that multiplicative inverses exist, the integers modulo p form a structure called a *field*. When p is composite, so that inverses may not exist and division may not be defined, all the other main algebraic laws still hold and we have a *ring*. There are many different structures with similar properties, so these concepts came into widespread use in algebra.

Under addition, **GF**(p) behaves remarkably like a group of transformations, except that its elements are not transformations. If we interpret the elements of **GF**(5) so that g corresponds to rotation of the plane about the origin through an angle $2\pi g/5$, then addition in **GF**(p) corresponds exactly to addition of angles. For instance, $4\oplus1$ corresponds to $8\pi/5 + 2\pi/5 = 10\pi/5 = 2\pi$, the same angle as 0. So these two structures are identical except for the context in which they are defined. They are said to be *isomorphic*.

There is another structure in **GF**(5) that closely resembles a group of transformations: its nonzero elements under multiplication. Now there are four elements, and they form a group isomorphic to the rotational symmetries of the square. In this sense, a Galois field is two groups joined together: a group of nonzero elements under multiplication, and a larger group with 0 included, under addition. The distributive law places a constraint on how the two groups relate to each other.

If $n > 1$, the integers modulo n do not form a field, because $p.p^{n-1} \equiv 0$ (mod p^n). The definition of **GF**(p^n) is more complicated.

Elliptic functions

Groups turned up in complex analysis because they brought together several different strands of the subject, unifying them into what is now a very powerful toolkit, with applications to other areas including number theory and algebraic geometry. It played a key role in Andrew Wiles's 1995 proof of Fermat's Last Theorem, for example.

In real analysis, the trigonometric functions 'sine' and 'cosine' are highly influential. One of their important properties, already noted in the context of waves, is *periodicity*. Their values remain the same if 2π is added to the variable:

$$\sin(x + 2\pi) = \sin(x) \qquad \cos(x + 2\pi) = \cos(x)$$

It immediately follows that adding an integer multiple $2k\pi$ to x also leaves the function unchanged. This relationship also holds when the variable is complex (replace x by $z = x + iy$). Another closely related periodic function on the complex plane is the exponential e^x, but this time the period is $2\pi i$, which is imaginary. The link is the famous equation:

$$e^{i\theta} = \cos \theta + i \sin \theta$$

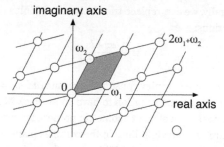

11. The lattice formed by all-integer linear combinations of two complex periods ω_1 and ω_2

Because the complex numbers form a plane, it is seems possible in principle for a complex function f to have two *independent* periods, ω_1 and ω_2, so that:

$$f(z + \omega_1) = f(z + \omega_2) = f(z)$$

By 'independent' I mean that ω_1 is not a real multiple of ω_2, so ω_1 and ω_2 correspond to linearly independent vectors in the real plane. The linear combinations $m\omega_1 + n\omega_2$, for integer m, n, form a lattice; see Figure 11. The function is completely determined by its values on any 'tile' of this lattice, such as the shaded region. The values elsewhere are obtained by translating this tile by an element of the lattice. Specifically, the equation

$$f(z + m\omega_1 + n\omega_2) = f(z)$$

defines the values of f on the shaded tile translated by $m\omega_1 + n\omega_2$.

Functions of this kind are called *elliptic functions*. The name reflects the historical path to their discovery: they arise when calculating the length of an arc of an ellipse. 'Doubly periodic function' is a more informative name. Elliptic functions can be constructed using infinite series that sum certain expressions over the lattice.

More generally, we can replace translations by Möbius transformations:

$$z \rightarrow \frac{az+b}{cz+d}$$

where a, b, c, d are complex constants such that $ad-bc \neq 0$ (the condition for the transformation to have an inverse). Möbius transformations have elegant geometric properties; in particular they map circles or straight lines in the complex plane to circles or straight lines. Composing two Möbius transformations gives another Möbius transformation, and the numbers a, b, c, d behave exactly like 2×2 matrices

$$\begin{bmatrix} a & b \\ c & d \end{bmatrix}$$

under matrix multiplication, though you get the same Möbius transformation if all four numbers are multiplied by the same constant, and this must be borne in mind.

Elliptic functions are invariant under a group of translations of the complex plane. Analogous *elliptic modular functions* are invariant under a suitable group of Möbius transformations. There are several standard ways to visualize such groups. One is to see what they do to the unit disc $|z| \leq 1$. Figure 12 shows a tiling of the unit disc whose symmetries comprise one particular group of Möbius transformations. Although the tiles seem to shrink towards the edge of the disc, they are all the same size in the metric—notion of distance—of the hyperbolic plane.

The unit disc is a standard model for one kind of non-Euclidean geometry: hyperbolic geometry, in which parallels (to a given line, passing through a given point) are not unique. In this model, 'straight lines' correspond to circles that cut the boundary of the disc at right angles. Möbius transformations turn out to be the analogue of rigid motions in this model of hyperbolic geometry.

12. Tiling of the unit disc corresponding to a group of Möbius transformations

This identification of Möbius geometry with hyperbolic geometry is one example of Klein's unification of the vast range of geometries that was proliferating late in the 19th century. His Erlangen Programme, named after the city where he announced it, takes each kind of geometry and associates to it the group of allowable transformations. For Euclidean geometry this is the group of rigid motions; for hyperbolic geometry it is the analogous group in hyperbolic space; for Möbius geometry it is the group of Möbius transformations; for projective geometry it is the group of projective transformations; and for topology it is the group of all continuous invertible transformations. If two apparently distinct geometries have isomorphic groups—more precisely, groups whose actions on their spaces are isomorphic— then they are really the same geometry in disguise. Geometry is then the study of invariants of transformation groups—features of the underlying space that are preserved by the transformations. At the time, this was an important way to unify geometry, and it inspired some profound new ideas.

Knot theory and topology

Topology is a kind of geometry, but instead of just rigid motions, any invertible continuous deformation is allowed. Rigid motions preserve features like lengths and angles, but continuous deformations do not—they can bend things, stretch them, or shrink them. A triangle can be changed into a circle by a continuous deformation. One of the founding papers in topology is 'Analysis situs', published in 1895 by Poincaré. It introduced a structure known as the *fundamental group*, associated with any topological space. Two spaces that are continuously deformable into each other have isomorphic fundamental groups; that is, the fundamental group is a topological invariant.

The fundamental group is defined using closed loops inside the topological space. First, choose a base point—any specific point in the space. Then consider all possible *loops*: continuous curves that start from the base point, wander around the space, and end back at the base point. Any two loops can be combined: first trace one, then the other. The trivial loop (stay at the base point) acts *almost* like an identity element. Reversing a loop by tracing it in the opposite direction is *almost* an inverse for the original loop. However, this doesn't quite work: tracing a loop and then going back along it is not the same as staying at the base point the whole time.

These deficiencies can be remedied by considering not loops, but *homotopy classes* of loops. Two loops are said to be *homotopic* if each can be deformed continuously into the other. Homotopy classes can be combined by combining representative loops and taking the class of all loops homotopic to the result. Now the homotopy class of the trivial loop really is an identity, and the homotopy class of the reverse loop is the inverse. In other words, loops form a group in which two loops are combined by tracing them in turn, provided the loops are considered only 'up to homotopy': homotopic loops are considered to be the same.

For example, suppose that the space is a circle. Then each homotopy class corresponds to all loops with a given *winding number*: how many times the loop winds round the circle in a clockwise direction. If you combine a loop with winding number m and one with winding number n, the result has winding number $m + n$. So the fundamental group of the circle corresponds precisely to integers under addition. The trivial loop has winding number 0; the reverse of a loop with winding number n has winding number $-n$.

Kurt Reidemeister took up Poincaré's idea, and used it to study knots. A knot is a closed curve K embedded in three-dimensional space (3-space). As a topological space in its own right, K is just a circle. What matters for knots is how K sits inside 3-space. One way to describe this embedding is to consider the *knot complement*: everything in 3-space that is *not* in K. Reidemeister defined the knot group of K to be the fundamental group of the knot complement. It is a topological invariant for the way K sits inside 3-space.

Reidemeister realized that it is possible to compute the knot group from a diagram of the knot by associating symbols $x_1, \ldots, x_n$ with a diagram of the knot. These symbols belong to the group, which implies that the group must also include all 'words' formed by the symbols, such as $x_1^2 x_2^{-3} x_1 x_3^{-5}$. To compose two words, write them one after the other and simplify if necessary. To represent the topology accurately, some words must be considered to be equivalent. The symbols are required to satisfy particular algebraic relations that encode the topology of the knot complement.

Figure 13 shows an example using a so-called Wirtinger presentation. On the left is a knot diagram, which naturally decomposes into connected arcs. Each arc is given a symbol. At each crossing the arrows look like either Figure 13 (middle) or Figure 13 (right). Using x, y, z to represent the corresponding symbols, we impose the relation $xy = zx$ for Figure 13 (middle), and $xz = yx$ for Figure 13 (right). These relations are geometric consequences of systematic ways to deform specific loops near crossings.

Here the elements of the group are not transformations. They are strings of symbols, with rules that require superficially different

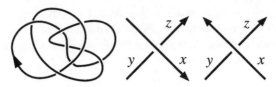

13. *Left*: A typical knot diagram broken into arcs at crossings. *Middle*: Three symbols at a crossing for one orientation. *Right*: Three symbols at a crossing for the other orientation

strings to be equal. Geometrically, they represent homotopy classes of loops, but their topological features have been distilled into a purely symbolic form.

Abstract groups

Ironically, this very richness of sources obscured the underlying simplicity of the group concept, because mathematicians and physicists defined their notion of a group in terms of the area in which they were using it. Many came close to the modern definition, but omitted crucial features—for example, the presence of the identity transformation. Today's definition seems to have evolved, gradually, from many closely related variants.

In an axiomatic approach, a group is like a group of transformations, except that you throw away the transformations. The elements (or members) of the group can in principle be anything. What matters is how they combine. Here is (one version of) the current definition:

A group is a set G together with an operation * that combines any two elements g and h of G to give an element $g*h$ of G. (Technically this is a function $*:G \times G \to G$.) The following conditions must hold:

1. *Identity*. There exists a special element in G, which we denote by 1, such that $1*g = g$ and $g*1 = g$ for all $g \in G$. (It need not be the *number* 1.)

2. *Inverse.* For any $g \in G$ there exists $g^{-1} \in G$ such that $g*g^{-1} = 1$ and $g^{-1}*g = 1$.

3. *Associative law.* For any $g, h, k \in G$ we have $g*(h*k) = (g*h)*k$.

Symmetry groups of geometric figures and groups of permutations satisfy this definition. The operation * is then composition of transformations, 1 is the identity transformation, g^{-1} is the inverse transformation, and the associative law holds because it holds for all functions whenever composition is defined. Similar remarks apply to groups of symbols, but now * is 'juxtapose and simplify'. The same goes for all of the other group-like structures that mathematicians had been inventing.

In particular, the integers modulo n form a group under the operation of addition, with identity 0 and inverse $-g$. When p is prime, the nonzero integers modulo p form a group under the operation of multiplication, with identity 1 and inverse $1/g$ modulo p.

There is a further condition, valid in some groups but not others:

4. *Commutative law.* For any $g, h \in G$ we have $g*h = h*g$.

A group with this property is said to be *abelian* (in memory of Abel) or *commutative*. The integers modulo n under addition, and the nonzero integers modulo a prime p under multiplication, are both abelian. By convention, the operation in an abelian group is often denoted by +, the identity by 0, and the inverse of g by $-g$. Sometimes this convention would cause confusion, as it would with the second of those two groups. If so, it is abandoned.

Finally we mention a useful technical term. The *order* of a group is the number of elements that it contains. This may be finite or infinite, depending on the group.

Chapter 3
Types of symmetry

Rigid motions are among the easiest symmetries to understand, because they have a geometric interpretation and their effects can be illustrated using pictures. The possibilities depend on the dimension of the space: the greater the dimension, the more different kinds of rigid motion there are.

On a line, there are two types of rigid motion. Either the orientation of the line (the direction in which the coordinate increases from negative to positive) is preserved, or not. If it is preserved, then the entire line is translated by some amount a, so a general point x maps to $x + a$. If not, then the line is reflected in the origin and then translated, so x maps to $-x + a$.

The possibilities become richer when we consider rigid motions in the plane. The main types, shown in Figure 14, are:

1. *Translations*, which move the entire plane in some direction by a specific distance.
2. *Rotations*, which rotate the plane through some angle about a fixed point.
3. *Reflections*, which map each point to its mirror image in some fixed line.

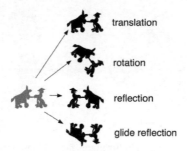

translation

rotation

reflection

glide reflection

14. The four types of rigid motion in the plane

Less well known, but also important, are:

4. *Glide reflections*, which map each point to its mirror image in
    some fixed line, and then translate the plane in the direction of
    that line.

The rigid-motion symmetries of a *bounded* region of the plane
cannot include a nontrivial translation or glide reflection, because
repeated application of either of these transformations moves
points through arbitrarily large distances. So for bounded shapes,
only rotations and reflections occur.

Cyclic and dihedral groups

The finite groups of rigid motions fall into two classes, depending
on whether the group consists only of rotations, or includes at
least one reflection.

Figure 15 shows two typical cases. The left-hand shape is
symmetric under five rotations about its centre, through angles
of 0, $2\pi/5$, $4\pi/5$, $6\pi/5$, and $8\pi/5$; see Figure 16 (left). These
rotations form the *cyclic group* of order 5, denoted by $\mathbf{Z}_5$. The
right-hand shape is symmetric under the same five rotations, but

15. Two symmetric shapes in the plane. *Left*: Z_5 symmetry. *Right*: D_5 symmetry

it also has five reflectional symmetries, in the mirror lines shown in Figure 16 (right). These rotations and reflections form the *dihedral group* of order 10, denoted by $\mathbf{D}_5$. (Many books use $\mathbf{D}_{10}$ instead, but the notation $\mathbf{D}_5$ reminds us of the relation to $\mathbf{Z}_5$.)

Similarly we can define the *cyclic group* $\mathbf{Z}_n$ *of order* n and the *dihedral group* $\mathbf{D}_n$ *of order* $2n$. The group $\mathbf{Z}_n$ consists of all rotations through $2k\pi/n$ about the origin, where $0 \leq k \leq n-1$. The group $\mathbf{D}_n$ consists of the same rotations, together with reflections in mirror lines making angles $k\pi/n$ with the

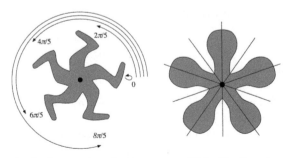

16. *Left*: The five rotations for Z_5 symmetry. *Right*: The five mirror lines for the extra symmetries in D_5

horizontal axis, where again $0 \le k \le n-1$. The dihedral group $\mathbf{D}_n$ is the symmetry group of a regular n-gon, and the cyclic group $\mathbf{Z}_n$ is the group of rotational symmetries of the n-gon. Here $\mathbf{Z}_n$ is a subset of $\mathbf{D}_n$ that happens to form a group under the same operation. We call it a *subgroup*.

Every finite group G of rigid motions of the plane must fix some point. In fact, if x is any point in the plane then a short calculation proves that the 'centre of mass'

$$\frac{1}{|G|} \sum_{g \in G} g(x)$$

is fixed by G. It is not hard to prove that $\mathbf{Z}_n$ $(n \ge 1)$ and $\mathbf{D}_n$ $(n \ge 1)$ comprise all finite groups of rigid motions of the plane that fix the origin.

Orthogonal and special orthogonal groups

There are two further important groups of rigid motions that fix the origin: the group $\mathbf{SO}(2)$ of all rotations about the origin and the group $\mathbf{O}(2)$ of all rotations and reflections in lines through the origin. The symbols stand for 'special orthogonal group' and 'orthogonal group'. A circle has symmetry group $\mathbf{O}(2)$. Again, $\mathbf{SO}(2)$ is a subgroup of $\mathbf{O}(2)$.

Friezes

Unbounded shapes can have a richer range of symmetries. A *frieze pattern* is a pattern in the plane whose symmetries leave the horizontal axis invariant. Individual points on that axis may move, but the whole axis is mapped to itself as a set. The name comes from the friezes often used at the top of, or across the middle of, wallpaper. There are seven different symmetry types of frieze, shown in Figure 17.

17. The seven symmetry types of frieze pattern

46

Wallpaper

Wallpaper patterns are symmetric under two independent translations: one step along the length of the roll of paper, and one step sideways to the next strip of paper, possibly with a shift up or down (which interior decorators call the 'drop'). This is just like the symmetries of elliptic functions, defined by a lattice; see Chapter 2. In addition, the entire pattern may be symmetric under various rotations and reflections. The simplest such symmetry group consists of combinations of the two translations, and a typical pattern is shown in Figure 18 (left). To avoid confusion, it is important to appreciate that the pattern in Figure 18 (right) does *not* have additional rotational symmetries. A single star has extra symetries, but when you apply them to the whole pattern, other stars don't map correctly. However, the pattern is now symmetric under reflection in the vertical bisector of any star, a new symmetry that the donkey paper lacks.

In 1891 the Russian pioneer of mathematical crystallography Yevgraf Fyodorov (often spelt Evgraf Fedorov) proved that there are exactly seventeen different symmetry classes of wallpaper pattern. George Pólya obtained the same result independently in

18. *Left*: **Wallpaper pattern with two independent translations (arrows).** *Right*: **Same translations, *no* $2\pi/5$ rotations, but also reflections in a vertical bisector of any pentagon (such as the dotted line)**

1924. These patterns can be classified according to the symmetry type of the underlying plane lattice. Replacing the points of the lattice by symmetrically arranged shapes produces patterns whose symmetry group is a subgroup of that of the lattice.

Lattices have two distinct types of symmetry: the lattice translations themselves, and the *holohedry* group, consisting of all rigid motions that fix one lattice point (which we may take to be the origin) and map the lattice to itself. Any symmetry is a combination of these two types. The proof is simple: suppose s is a symmetry of the lattice, sending 0 to $s(0)$. There is a translation t that maps 0 to $s(0)$, so $t^{-1}s = h$ is in the holohedry group. Therefore $s = th$, which is a holohedry composed with a lattice translation.

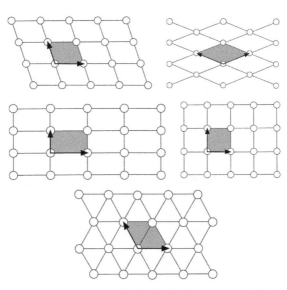

19. The five symmetry types of lattice in the plane. *Left to right*: **Parallelogrammic, rhombic, rectangular, square, hexagonal. Arrows show a choice of lattice generators. Shaded area is the associated fundamental region**

The five symmetry types of lattice, shown in Figure 19, are:

1. *Parallelogrammic* or *oblique*: the lattice generators are of unequal length and not at right angles. The fundamental domain is a parallelogram. The holohedry is $\mathbf{Z}_2$ generated by rotation through π.

2. *Rhombic*: the lattice generators are of equal length and not at angles $\pi/2$, $\pi/3$, $2\pi/3$. The fundamental domain is a rhombus. The holohedry is $\mathbf{D}_2$ generated by rotation through π and a reflection.

3. *Rectangular*: the lattice generators are of unequal length and at right angles. The fundamental domain is a rectangle. The holohedry is also $\mathbf{D}_2$.

4. *Square*: the lattice generators are of equal length and at right angles. The fundamental domain is a square. The holohedry is $\mathbf{D}_4$ generated by rotation through $\pi/2$ and a reflection.

5. *Hexagonal* or *(equilateral) triangular*: the lattice generators are of equal length and at an angle of $\pi/3$. The fundamental domain is a rhombus composed of two equilateral triangles. Parts of three of these fit together to make a hexagon that also tiles the plane. The holohedry is $\mathbf{D}_6$ generated by rotation through $\pi/6$ and a reflection.

To obtain the wallpaper classification, we examine each of these five types of lattice, and classify the subgroups of the symmetry group that contain the lattice translations. Figure 20 shows all seventeen wallpaper patterns, their standard crystallographic notation, and the underlying lattices.

Regular solids

Now we move up to three dimensions. A solid (that is, a polyhedron) is said to be *regular* if its faces are regular polygons, all identical, with the same arrangement of faces at each vertex. The five regular solids shown in Figure 21 are a rich source of symmetries in three dimensions. They are:

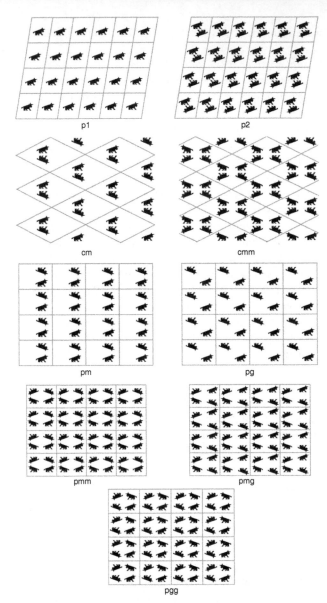

20. The seventeen wallpaper patterns. Captions are standard crystallographic notation

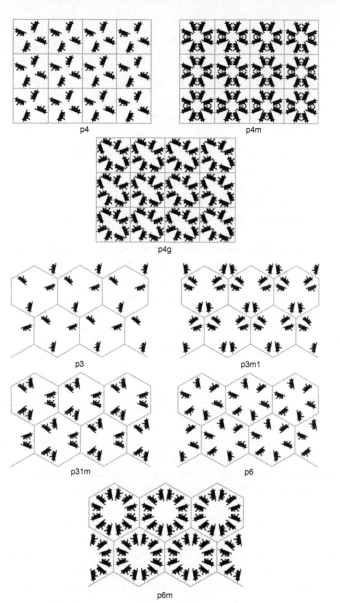

p4

p4m

p4g

p3

p3m1

p31m

p6

p6m

20. Continued

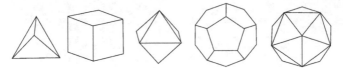

21. The five regular solids. *Left to right*: **Tetrahedron, cube, octahedron, dodecahedron, icosahedron**

- *Tetrahedron*: four faces, each an equilateral triangle.
- *Cube*: six faces, each a square.
- *Octahedron*: eight faces, each an equilateral triangle.
- *Dodecahedron*: twelve faces, each a regular pentagon.
- *Icosahedron*: twenty faces, each an equilateral triangle.

Not only are the faces and vertex arrangements of these solids very regular: the entire solid is, in the sense of symmetry. For each solid, any face can be mapped to any other face by a rigid motion, and this rigid motion maps the entire solid to itself. Moreover, any symmetry of that face extends uniquely to a symmetry of the solid. Proving these statements is not especially hard, but it requires developing a few geometric techniques not found in Euclid.

These symmetry properties allow us to compute the number of symmetries that each regular solid possesses; that is, the order of its symmetry group. For example, the tetrahedron has four faces, each of which can be mapped to a specified reference face. Then the reference face has six symmetries—the group $\mathbf{D}_3$—all of which extend to the whole solid. So in total there are 4.6 = 24 symmetries. More generally, if the solid has F faces, each with E edges, its symmetry group contains $2EF$ rigid motions. Table 1 shows the results.

Looking at this table, it immediately becomes clear that the cube and octahedron have the same number of symmetries, as do the dodecahedron and icosahedron. There is a simple reason, known as duality. The centres of the faces of a cube form the vertices of an octahedron, so any symmetry of the cube is also a symmetry of

Table 1. Number of symmetries of regular solids

Solid	F	E	Number of symmetries ($= 2EF$)
tetrahedron	4	3	24
cube	6	4	48
octahedron	8	3	48
dodecahedron	12	5	120
icosahedron	20	3	120

this octahedron. On the other hand, the centres of the faces of an octahedron form the vertices of a cube, so any symmetry of the octahedron is also a symmetry of this cube. A similar relationship holds for the dodecahedron and icosahedron. So these pairs of symmetry groups are isomorphic.

What about the humble tetrahedron? The centres of its faces form another tetrahedron. It is self-dual, and this construction gives rise to nothing new.

A symmetry of a regular solid always fixes its centre, which by convention is the origin. Suppose we define an orientation for the solid by conceptually marking an arrow on each face in an anticlockwise direction, as seen from outside the solid. Rotations preserve this orientation. Reflections, and some other transformations, reverse it. In fact, the symmetry preserves the orientation of the solid if and only if it is a rotation in three-dimensional space. It reverses the orientation if and only if it is a rotation composed with minus the identity. This sends each vertex to the diametrically opposite one, mapping (x,y,z) to $(-x,-y,-z)$, and it can be written as $-I$.

Reflections can be characterized by two simple properties: they fix every point in a plane through the origin, the *mirror plane*, and

53

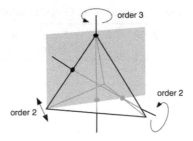

order 3

order 2

order 2

22. Symmetries of a tetrahedron

they act as minus the identity on the line normal to that plane. Of the $2EF$ symmetries of the solid, EF are rotations and the remaining EF are rotations composed with a reflection or $-I$. In general, reflections alone do not give all the orientation-reversing symmetries of three-dimensional space. For example, $-I$ is a symmetry of the cube. Although this map reverses orientation, it is not a reflection because the only point that it fixes is the origin.

The regular solids therefore provide three symmetry groups: the tetrahedral group **T**, the octahedral group **O** (which also corresponds to the cube), and the icosahedral group **I** (which also corresponds to the dodecahedron and is often called the dodecahedral group). We consider the simplest case, the tetrahedron, to see how the various rigid motions act; see Figure 22.

Tetrahedral group

There are five geometrically distinct kinds of symmetry, summarized in Table 2:

- The identity. This is (trivially) a rotation that fixes every point. There is one such transformation.
- Rotations fixing a vertex: two for each vertex. Each such rotation has order 3. There are eight of them.
- Rotations about an axis joining the midpoints of opposite sides. Each such rotation has order 2. There are three of them.

Table 2. Types of symmetry for tetrahedron

Motion	Order	Number
identity	1	1
vertex rotation	3	8
mid-axis rotation	2	3
reflection	2	6
rotate and reflect	4	6

- Reflections in a plane passing through two vertices and the midpoint of the opposite edge. Each has order 2. There are six of them.
- Motions that cycle the four vertices in some order, fixing none of them. Geometrically such a motion can be obtained by rotating through $\pi/2$ about an axis joining the midpoints of opposite edges, and then reflecting the tetrahedron in a plane at right angles to that axis. There are six such motions (rotate clockwise or anticlockwise for each of three axes) and each has order 4.

Note that $-I$ does not leave a tetrahedron invariant.

In the case of the octahedral and icosahedral groups, we describe only the orientation-preserving motions (rotations) for simplicity. The orientation-reversing motions can be obtained by composing these with $-I$. Some are reflections, some are not.

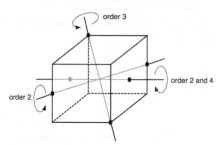

23. Rotational symmetries of a cube

Octahedral group

It is easiest to visualize this using a cube; see Figure 23. Now we get Table 3:

- The identity. There is one such transformation.
- Rotations about an axis joining the midpoints of opposite sides. Each such rotation has order 2. There are six of them.
- Rotations fixing a vertex: two for each vertex. Each such rotation has order 3. There are eight of them.
- Rotations by $\pm\pi/2$ about an axis joining the midpoints of opposite faces. Each such rotation has order 4. There are six of them.
- Rotations by π about an axis joining the midpoints of opposite faces. Each such rotation has order 2. There are three of them.

Table 3. Rotational symmetries of the cube

Motion	Order	Number
identity	1	1
mid-axis rotation	2	6
vertex rotation	3	8
mid-face rotation $\pm\pi/2$	4	6
mid-face rotation π	2	3

Icosahedral group

It is easier to draw pictures using a dodecahedron; see Figure 24. Now we get Table 4:

- The identity. There is one such transformation.
- Rotations $\pm2\pi/5$ about an axis joining the midpoints of opposite faces. Each such rotation has order 5. There are twelve of them.
- Rotations $\pm4\pi/5$ about an axis joining the midpoints of opposite faces. Each such rotation has order 5. There are twelve of them.

- Rotations fixing a vertex: two for each vertex. Each such rotation has order 3. There are twenty of them.
- Rotation by π about an axis passing through the midpoints of opposite edges. Each such rotation has order 2. There are fifteen of them.

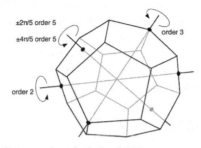

24. **Rotational symmetries of a dodecahedron**

Table 4. Rotational symmetries of the dodecahedron

Motion	Order	Number
identity	1	1
mid-face rotation $\pm 2\pi/5$	2	12
mid-face rotation $\pm 4\pi/5$	2	12
vertex rotation	3	20
mid-edge rotation	2	15

Orthogonal group

Several symmetry groups in three-dimensional space contain infinitely many transformations. If we fix an axis, then all rotations about that axis form a group isomorphic to **SO**(2), and reflections in planes through that axis extend this to a group isomorphic to **O**(2). The cone (or any 'solid of revolution') is an example of this type of symmetry. A cylinder has another type of symmetry as well: reflection about a plane at right angles to its axis.

If we include all possible rotations about all possible axes, we get the special orthogonal group **SO**(3). Augmented by all rotations composed with $-I$, this gives the orthogonal group **O**(3). The obvious shape with **O**(3) symmetry is a sphere. If a shape in three dimensions has **SO**(3) symmetry then it must also have **O**(3) symmetry, so we have to add some extra structure to get **SO**(3). For instance, we can provide the sphere with an orientation and require the transformation to preserve this.

Crystallographic groups

The regular shapes of crystals can be traced to the arrangement of their atoms, which in an ideal model form a regular lattice, symmetric under three independent translations in three-dimensional space. So a crystal is the three-dimensional analogue of wallpaper. Several different classifications of crystal lattices are possible, providing increasing levels of fine detail. The coarsest classification lists the lattices in terms of their symmetries: these are called Bravais lattices or lattice systems. They are listed in Table 5 and illustrated in Figure 25.

We saw in Figure 19 that in two dimensions the analogous list contains five types of lattice, but Figure 20 shows seventeen symmetry classes of patterns. The same distinction arises in three dimensions: Bravais lattices classify the symmetry types of *points* arranged in a lattice, whereas the full list classifies *shapes* arranged in a lattice. The shapes have a richer range of symmetries, and distinguish more types of pattern. The most extensive list in three dimensions classifies *space groups*: symmetry groups of three-dimensional arranged on a lattice. There are 230 of these, or 219 if certain mirror-image pairs are considered to be equivalent.

Table 5. The fourteen Bravais lattices: names

Number	Lattice system	Name
1	triclinic	triclinic
2	monoclinic	monoclinic
3	monoclinic	monoclinic base-centred
4	orthorhombic	orthorhombic
5	orthorhombic	orthorhombic base-centred
6	orthorhombic	orthorhombic body-centred
7	orthorhombic	orthorhombic face-centred
8	tetragonal	tetragonal
9	tetragonal	tetragonal body-centred
10	rhombohedral	rhombohedral
11	hexagonal	hexagonal
12	cubic	cubic
13	cubic	cubic body-centred
14	cubic	cubic face-centred

A curious feature observed in these classifications is the *crystallographic restriction*: a crystal lattice in two or three dimensions cannot have a symmetry of order 5. In fact, the only permissible orders are 1, 2, 3, 4, and 6. Here is a simple proof of this fact for lattices in the plane. First, note that every lattice is *discrete*: the distance between distinct points always exceeds some nonzero lower bound. This is intuitively clear; the proof is a simple estimate. Suppose the lattice consists of all integer linear combinations $au + bv$ of two vectors u and v. We can choose coordinates so that $u = (1,0)$, in which case $v = (x,y)$ with $y \neq 0$ because v is

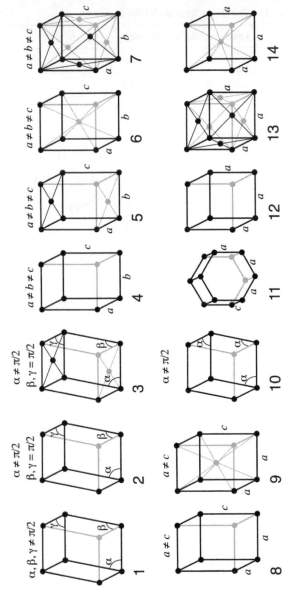

25. **The fourteen Bravais lattices: geometry**

independent of u. Suppose that $au + bv$ is not the origin $(0,0)$. The square of the distance from $au + bv$ to the origin is:

$$|| (a+bx, by) ||^2 = (a+bx)^2 + (by)^2 \geq b^2 y^2 \geq y^2$$

unless $b = 0$, since b is an integer. But then:

$$|| (a+bx, by) ||^2 = a^2 \geq 1$$

since a is nonzero in this case. Therefore the distance from the origin to any other lattice point is at least $\min(1, y^2)$, a fixed constant greater than 0. By translation, the same bound holds for the distance between any two distinct lattice points.

Now suppose that a lattice has a point X with order-5 symmetry. The symmetry must be a rotation since reflections have order 2. The point X may lie in the lattice, but conceivably it may not: a square lattice has 90° rotational symmetry about the centre of any square, which is not in the lattice, for example. No matter: pick any lattice point A that differs from X, and use the order-5 symmetry repeatedly to rotate it into successive positions B, C, D, E. These must be in the lattice because we are using a symmetry.

Now we have a regular pentagon ABCDE consisting of lattice points; see Figure 26 (left). Fill in the five-pointed star to find

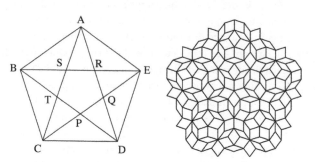

26. *Left*: Pentagon and five-pointed star. *Right*: **Part of Penrose pattern with order-5 symmetry**

points P, Q, R, S, T as shown. ABPE is a parallelogram, indeed a rhombus. The vector BP is equal to the vector AE, which is a lattice translation. Therefore P lies in the lattice. Similarly Q, R, S, and T lie in the lattice. We have now found a *smaller* regular pentagon whose vertices all lie in the lattice. In fact, its size is

$$\frac{3-\sqrt{5}}{2} \approx 0.382$$

times that of the original pentagon. By repeating this construction, the distance between two distinct lattice points can be made arbitrarily small; however, this is impossible. Contradiction.

In four dimensions there are lattices with order-5 symmetries, and any given order is possible for lattices of sufficiently high dimension. You might like to consider adapting the above proof to three dimensions, and then working out why it fails in four.

Although order-5 symmetries of a crystal lattice do not exist in two or three dimensions, Roger Penrose (inspired by Johannes Kepler) discovered non-repeating patterns in the plane with a generalized type of order-5 symmetry. They are called *quasicrystals*. Figure 26 (right) is one of two quasicrystal patterns with exact fivefold symmetry. In 1984 Daniel Schechtman discovered that quasicrystals occur in an alloy of aluminium and manganese. Initially most crystallographers discounted this suggestion, but it turned out to be correct, and in 2010 Schechtman was awarded the Nobel Prize in Chemistry. In 2009, Luca Bindi and his colleagues found quasicrystals in an alloy of aluminium, copper, and iron: mineral samples from the Koryak mountains in Russia. To find out how these quasicrystals formed, they used mass spectrometry to measure the proportions of different isotopes of oxygen. The results indicate that the mineral is not of this world: it derives from carbonaceous chondrite meteorites, originating in the asteroid belt.

Permutation groups

We now move on to a class of groups that does not come from geometry. A *permutation* on a set X is a map $\rho : X \to X$ that is one-to-one and onto, so that the inverse ρ^{-1} exists. Intuitively, ρ is a way to rearrange the elements of X. For example, suppose that $X = \{1, 2, 3, 4, 5\}$ and $\rho(1) = 2$, $\rho(2) = 3$, $\rho(3) = 4$ $\rho(4) = 5$, $\rho(5) = 1$. Then ρ rearranges the ordered list $(1, 2, 3, 4, 5)$ to give $(2, 3, 4, 5, 1)$. The notation

$$\rho = \begin{pmatrix} 1 & 2 & 3 & 4 & 5 \\ 2 & 3 & 4 & 5 & 1 \end{pmatrix}$$

shows this clearly. Diagrammatically, the effect of ρ is shown in Figure 27 (left), and in an alternative form in Figure 27 (right).

Let $X = \{1, 2, 3, \ldots, n\}$ where n is a positive integer. The set of all permutations of X is a group under composition. The identity map is a permutation, the inverse of a permutation is a permutation, and $(fg)^{-1} = g^{-1}f^{-1}$ so the composition of two permutations is a permutation. This group is the *symmetric group* $\mathbf{S}_n$ on n symbols. Its order is $|\mathbf{S}_n| = n!$

In Figure 27 (left) the long arrow crosses the other four arrows. We write $c(\rho) = 4$, where c is the *crossing number*, defined to be the smallest number of crossings in such a

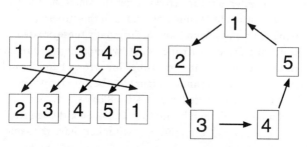

27. **Two ways to illustrate the effect of the permutation ρ**

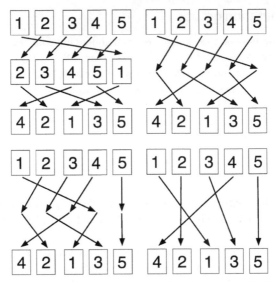

28. **Composing two permutations.** *Top left*: Composition. *Top right*: Removing middle layer. *Bottom left*: One stage in straightening out the arrows from 5 and 1. *Bottom right*: All arrows straightened

diagram. Suppose we compose this permutation ρ with another permutation σ, obtaining $\sigma\rho$ as in Figure 28 (top left). Removing the middle layer, we see there are $c(\rho) + c(\sigma)$ crossings. However, this is not the smallest number of crossings for $\sigma\rho$ because some arrows cross each other more than once. We can straighten out the arrows to get the smallest number. Figure 28 (bottom left) shows one stage in this process, involving the arrows from 1 and 5. Originally these crossed twice; moving the arrows cancels out two crossings and reduces this to zero. Another pair of crossings for arrows 1 and 3 can be removed in the same manner. Figure 28 (bottom right) shows the final result. We started with $4 + 4 = 8$ crossings, then reduced it by 2 to 6, then by another 2 to 4. So although $c(\sigma\rho)$ is not the same as $c(\rho) + c(\sigma)$, these two numbers have the same *parity*: odd or even.

The same argument, which can be performed in a formal algebraic manner for rigour, shows that in general:

$$c(\sigma\rho) \equiv c(\rho) + c(\sigma) \pmod{2}$$

The value of $c(\rho)$ modulo 2 is called the *parity* of the permutation ρ. We say that is an *even* permutation if $c(\rho) \equiv 0$, and an *odd* permutation if $c(\rho) \equiv 1$. The equation implies that:

> even composed with even gives even;
> odd composed with odd gives even;
> even composed with odd gives odd;
> odd composed with even gives odd.

In particular, the set of all even permutations is a subgroup of $\mathbf{S}_n$. It is called the *alternating group* on n symbols, written $\mathbf{A}_n$. Its order is $|\mathbf{A}_n| = n!/2$.

There are many other groups of permutations. In fact, any group is isomorphic to some group of permutations.

A useful alternative notation for permutations decomposes them into cycles. A *cycle* is a permutation of distinct numbers $x_1, \ldots, x_m$ that sends x_j to x_{j+1} if $1 \le j \le m-1$, and sends x_m to x_1. We use the notation $(x_1 x_2 \ldots, x_m)$ and refer to this as an m-cycle. For example, the permutation ρ defined above is a *5-cycle*. It moves each number one place anticlockwise in Figure 27 (right). Every permutation can be written as the composition of disjoint cycles, that is, cycles with no numbers in common.

Chapter 4
Structure of groups

To provide techniques for analysing the structure of symmetry groups, and language to describe them, we now discuss some basic concepts of group theory. Formal proofs are generally omitted. This chapter is only the beginning of a vast theory of groups, introducing a few simple ideas that we will need in later chapters. It is of necessity more symbol-ridden and formal-looking than the rest of this book.

Isomorphism

We have already seen that sometimes two groups, which are technically distinct, may have the same abstract structure (we used the term 'isomorphic'). For example, the group $\mathbf{Z}_3$ of integers modulo 3 under addition has the group multiplication table:

	0	1	2
0	0	1	2
1	1	2	0
2	2	0	1

The rotational symmetries of an equilateral triangle form a group R with three elements, consisting of the rotations R_0, $R_{2\pi/3}$, $R_{4\pi/3}$. The multiplication table now looks like this:

	R_0	$R_{2\pi/3}$	$R_{4\pi/3}$
R_0	R_0	$R_{2\pi/3}$	$R_{4\pi/3}$
$R_{2\pi/3}$	$R_{2\pi/3}$	$R_{4\pi/3}$	R_0
$R_{4\pi/3}$	$R_{4\pi/3}$	R_0	$R_{2\pi/3}$

The two tables have exactly the same structure, but they use different symbols. If we take the first table and change 0 to R_0, 1 to $R_{2\pi/3}$, and 2 to $R_{4\pi/3}$ throughout, we obtain the second table. Formally, this feature is expressed in terms of the map $f\colon \mathbf{Z}_3 \to R$ defined by the change of symbols:

$$f(0) = R_0 \qquad f(1) = R_{2\pi/3} \qquad f(2) = R_{4\pi/3}$$

or, more succinctly:

$$f(j) = R_{2j\pi/3}$$

for $j = 0, 1, 2$. This map is a bijection, with the property:

$$f(j+k) = f(j)f(k)$$

This implies that the two tables have the same structure.

More generally, if G and H are groups, a map $f\colon G \to H$ is an *isomorphism* if it is a bijection and satisfies the condition:

$$f(gh) = f(g)f(h) \qquad \text{for all } g, h \in G$$

We say that G and H are *isomorphic*, and write $G \cong H$.

If two groups are isomorphic, then any property of the first that depends only on the abstract structure also holds for the second. In particular, they have the same order (recall that the *order* of a group is the number of elements that it contains). However, it is easy to find groups with the same order that are not isomorphic: for example, $\mathbf{Z}_6$ and $\mathbf{D}_3$. Both have order 6, but the first is abelian and the second is not.

Subgroup

We have already encountered several examples in which a group is contained in another group. The formal concept is defined like this: a *subgroup* of a group G is a subset $H \subseteq G$ such that:

1   $1 \in H$
2   If $h \in H$ then $h^{-1} \in H$
3   If $g, h \in H$ then $gh \in H$

Here are some examples of subgroups that we have already met:

- $\mathbf{Z}_n$ is a subgroup of $\mathbf{D}_n$.
- $\mathbf{A}_n$ is a subgroup of $\mathbf{S}_n$.
- $\mathbf{SO}(2)$ is a subgroup of $\mathbf{O}(2)$.
- $\mathbf{T}$, $\mathbf{O}$, $\mathbf{I}$ and $\mathbf{SO}(3)$ are subgroups of $\mathbf{O}(3)$.

If we realize the cyclic group $\mathbf{Z}_n$ as the integers modulo n under the operation of addition, we can be more ambitious and list all subgroups. It turns out that they correspond to integers m that divide n exactly. For each such m, there is a subgroup:

$$m\mathbf{Z}_n = \left\{ mj : j = 0, 1, \ldots, n/m - 1 \right\}$$

It is isomorphic to $\mathbf{Z}_{n/m}$. These are all of the subgroups of $\mathbf{Z}_n$.

For example, the divisors of 12 are 1, 2, 3, 4, 6, 12. So the subgroups of $\mathbf{Z}_{12}$ are:

$$1\mathbf{Z}_{12} = \mathbf{Z}_{12}$$
$$2\mathbf{Z}_{12} = \{0, 2, 4, 6, 8, 10\} \cong \mathbf{Z}_6$$

$$3\mathbf{Z}_{12} = \{0, 3, 6, 9\} \cong \mathbf{Z}_4$$
$$4\mathbf{Z}_{12} = \{0, 4, 8\} \cong \mathbf{Z}_3$$
$$6\mathbf{Z}_{12} = \{0, 6\} \cong \mathbf{Z}_2$$
$$12\mathbf{Z}_{12} = \{0\} \cong \mathbf{1}$$

It is no coincidence that the orders of these subgroups are divisors of 12, the order of the group. A similar property holds in complete generality.

Lagrange's Theorem

Let G be a finite group. If H is a subgroup of G then $|H|$ divides $|G|$.

For example, if G is the group of rotational symmetries of a dodecahedron then $|G| = 60$. So the possible orders of subgroups are 1, 2, 3, 4, 5, 6, 10, 12, 15, 20, 30, 60. Most of these orders occur, but not 15, 20, and 30.

Order of an element

Suppose that G is a group and $g \in G$. The set of all powers of g

$$H = \left\{ g^n : n \in Z \right\}$$

is a subgroup, the subgroup *generated by g*. There are two main possibilities:

- All powers of g are distinct. Then $H \cong \mathbf{Z}$, the additive group of integers. This case cannot occur if G is finite.
- Two distinct powers of g are equal. Then $H \cong \mathbf{Z}_k$ where k is the smallest positive integer such that $g^k = 1$. This case must occur if G is finite.

The *order* $|g|$ of g is ∞ in the first case and k in the second.

Since the powers of g form a subgroup, Lagrange's Theorem implies that when G is finite, the order of every element divides the order of G. For example, the orders of elements of the group of rotational symmetries of a dodecahedron must belong to the list given above for orders of subgroups. Table 5 shows that of these, only 1, 2, 3, 4, and 5 actually occur for elements.

Conjugacy

Two symmetries of some objects may be essentially the same, except that they are applied at different locations. For example, Figure 29 shows two reflections s and t of a regular pentagon, with different axes, and the rotation r that sends the axis of s to the axis of t.

It is clear from the picture, and can be checked using properties of $\mathbf{D}_5$, that:

$$t = r^{-1}sr$$

That is, to reflect in the first axis, rotate that axis to the second position, then reflect in that axis, then rotate back.

In general, if G is a group, then two elements $g, h \in G$ are said to be *conjugate* if there exists $k \in G$ such that $h = k^{-1}gk$. Conjugate elements always have the same order. The set of all elements conjugate to a given one is called a *conjugacy class*. For a finite

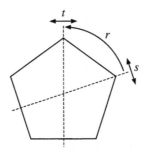

29. **Conjugate reflectional symmetries of a regular pentagon**

group G, the number of elements in any conjugacy class always divides the order of G. Informally, conjugate elements do the same thing, but in a different place.

Normal subgroups, homomorphisms, and quotient groups

Subgroups are the obvious way to derive smaller groups from a given one. However, there is a second, subtler construction, whose genesis can be traced to Galois's approach to the solution of equations by radicals, although not in its present formulation. This is called a quotient group, and it is fundamental to group theory.

For an example, consider the symmetry group of a square. The symmetries are of two kinds: rotations (which preserve orientation, that is, do not flip the square over) and reflections (which reverse orientation, that is, do flip the square over). The symbols *flip* and *non-flip* form a group in their own right:

$$non\text{-}flip \times non\text{-}flip = non\text{-}flip$$
$$non\text{-}flip \times flip = flip$$
$$flip \times non\text{-}flip = flip$$
$$flip \times flip = non\text{-}flip$$

That is, any symmetry of the first type, composed with any symmetry of the second type, always yields a symmetry of the third type. Abstractly, we recognize this group: it is cyclic of order 2.

For another example, consider the group $\mathbf{S}_n$ of all permutations of a set with n elements. Permutations have *parity*: they can be can be even or odd. The parity of a product of two permutations depends only on the types of the permutations themselves: *odd* × *even* = *odd* and so on. Again, these two types naturally form a group in their own right, and again it is cyclic of order 2.

A group constructed in this manner is called a *quotient group*. Formally, a quotient group can be defined as a special kind of partition of the group: a way to chop it into disjoint pieces. Suppose that this can be done in such a way that the pieces inherit the group structure. That is: if g_1 and g_2 are in the same piece, and h_1 and h_2 are in the same piece, then g_1h_1 and g_2h_2 are in the same piece. If this property holds, the set of pieces forms a group. In the above examples, the pieces are *flip* and *non-flip* for the symmetries of a square, and *even* and *odd* for permutations.

Intuitively, we can think of a quotient group as a way to colour the elements of the group so that elements have the same colour if and only if they lie in the same piece. The above condition implies that we can multiply any two colours together to obtain a well-defined colour. Choose two elements that have those colours, multiply them together, and take the colour of the result. The condition ensures that this product always has the *same* colour, whichever two elements we pick. The elements of the quotient group are now the colours, and the group operation is multiplication of colours.

We now have two groups: the original one G and a group of colours K. There is a natural map $\phi{:}G{\to}K$ in which $\phi(g)$ is the colour of $g \in G$. The ability to multiply colours consistently boils down to the equation

$$\phi(gh) = \phi(g)\,\phi(h)$$

for all $g, h \in G$. A map with this property is called a *homomorphism*. It is like an isomorphism, but need not be a bijection.

Although colourings of this kind are easy to visualize, it is not so easy to find them. An alternative characterization of quotient groups relates them to special subgroups, called *normal subgroups.* Whenever a group has a quotient group, the piece that contains the identity forms a subgroup. All elements of that piece have the same colour—suppose it is red. We claim that *red* × *red* = *red* in the quotient. This follows since $1 \times 1 = 1$, and 1 is red. This implies that the red piece of the group is closed under multiplication. Because $1^{-1} = 1$, it is also closed under the inverse. So it forms a subgroup.

Do all subgroups arise in this manner? It turns out that they do not. In fact, a subgroup H of G can be the piece containing 1, for some quotient group of G, if and only if H has one further property, called *normality*. This states that if h is any member of H, and g any member of G, then $g^{-1}hg$ is a member of H. This condition is necessary because in G we have $g^{-1}1g = g^{-1}g = 1$. It is also sufficient: define a partition of G by:

g_1 and g_2 belong to the same part if and only if $g_1 g_2^{-1}$ is a member of H.

The parts here are called *cosets* of H. A few calculations show that they have the required colouring property. The quotient group is denoted by G/H. There is a natural homomorphism $\phi : G \to G/H$, sending each element to its coset, and the elements of H map to the identity of G/H. So homomorphisms, colourings, and normal subgroups are three ways to describe the same idea.

The integers modulo n are quotient groups of $\mathbf{Z}$ under addition. This group is abelian, so every subgroup is normal. Consider for example the subgroup $N = 5\mathbf{Z}$, consisting of the multiples of 5. Its cosets are:

$N = \{5k : k \in \mathbf{Z}\}$
$N1 = \{5k + 1 : k \in \mathbf{Z}\}$
$N2 = \{5k + 2 : k \in \mathbf{Z}\}$
$N3 = \{5k + 3 : k \in \mathbf{Z}\}$
$N4 = \{5k + 4 : k \in \mathbf{Z}\}$

30. *Left*: Colouring integers modulo 5. *Right*: Addition table for the colours

and these exhaust the whole of **Z** because every number leaves remainder 0, 1, 2, 3, or 4 on division by 5. Figure 30 (left) shows the corresponding colouring, and Figure 30 (right) shows how the colours combine. More generally, for any integer $n \neq 0$ the quotient group $\mathbf{Z}/n\mathbf{Z}$ is the group $\mathbf{Z}_n$ of integers modulo n.

Chapter 5
Groups and games

Group theory has applications not just to science, but to games and puzzles. Here we look at three examples. The first is the Fifteen Puzzle, and group theory proves that it is impossible. The second is the Rubik cube, and group theory helps us to solve it. The third is sudoku: group theory tells us how many puzzles there are, but sheds little light on how to solve them.

In 1880 the USA, Canada, and Europe were swept by a short-lived craze—it started in April and was over by July—for a puzzle invented by a New York postmaster named Noyes Palmer Chapman. Matthias Rice, a businessman specializing in woodworking, marketed it as the Gem Puzzle; a dentist called Charles Pevey offered money for a solution. The puzzle is also known as the Boss Puzzle, Game of Fifteen, Mystic Square, and Fifteen Puzzle. It consists of fifteen sliding blocks, numbered 1–15, initially arranged as in Figure 31 (left), with an empty square at bottom right. The aim of the puzzle is to rearrange the blocks into Figure 31 (right), by sliding successive blocks into the empty square—which moves as the blocks are slid.

One hundred years later, a similar craze spread worldwide. But this time the puzzle consisted of moving cubes, rather than squares. It was the Rubik cube, Figure 32, invented by the Hungarian sculptor and architecture professor Ernö Rubik. To

1	2	3	4
5	6	7	8
9	10	11	12
13	15	14	

1	2	3	4
5	6	7	8
9	10	11	12
13	14	15	

31. Fifteen Puzzle. *Left*: **Start.** *Right*: **Finish**

32. A Rubik cube with one face in the process of being rotated

date over 350 million cubes have been sold. The six faces of the cube are coloured so that each face has one colour. The puzzle is constructed so that any face of the cube can be rotated, and a series of these rotations scrambles the colours. The aim of the puzzle is to restore the original colouring.

In 2005 yet another craze swept the world; this time it was a combinatorial puzzle whose solution required placing digits 1–9 in a 9×9 square divided into nine 3×3 subsquares, in such a manner that each row, column, and subsquare contained one of each digit. Some digits were filled in and the challenge was to complete the

5	3			7				
6			1	9	5			
	9	8					6	
8				6				3
4			8		3			1
7				2				6
	6					2	8	
			4	1	9			5
				8			7	9

5	3	4	6	7	8	9	1	2
6	7	2	1	9	5	3	4	8
1	9	8	3	4	2	5	6	7
8	5	9	7	6	1	4	2	3
4	2	6	8	5	3	7	9	1
7	1	3	9	2	4	8	5	6
9	6	1	5	3	7	2	8	4
2	8	7	4	1	9	6	3	5
3	4	5	2	8	6	1	7	9

33. *Left*: A sudoku grid. *Right*: Its solution

grid. This game is, of course, sudoku; see Figure 33. It remains wildly popular and features regularly in most newspapers, following in the well-trodden footsteps of crossword puzzles.

These puzzles all have a considerable amount of symmetry—more than meets the eye—and they illustrate how group theory illuminates structural symmetries in mathematics. We consider them in turn.

Fifteen Puzzle

The Fifteen Puzzle is often associated with the famous American puzzlist Sam Loyd, who claimed that he had started a craze for it in the 1870s. However, Loyd's contact with the puzzle was confined to writing about it in 1896. He revived interest by offering a prize of $1,000, a substantial sum at the time and still not to be sneezed at. But Loyd's money was safe, as he well knew. In 1879, William Johnson and William Story had proved that the Fifteen Puzzle is insoluble.

Their argument involves the 'potential' symmetry group of the puzzle, which consists of all possible permutations of sixteen objects—the fifteen sliding blocks and the empty square, which for

34. Colouring the squares in the Fifteen Puzzle

simplicity we label 16. It is therefore the symmetric group $\mathbf{S}_{16}$. This is a symmetry group in the sense that it comprises all possible ways to rearrange the blocks. However, the 'actual' symmetry group generated by all legal moves is a proper subgroup: not all of these arrangements can be realized using the prescribed moves.

Here's why. Sliding a block in effect transposes that block with the empty square, and this permutation is a 2-cycle. If we colour the squares like a chessboard, as in Figure 34, then each such move changes the colour associated with the empty square. So a sequence consisting of an even number of moves will leave the colour of the empty square unchanged, whereas a sequence consisting of an odd number of moves will change the colour of the empty square. The conditions of the puzzle require the empty square to end up in its original position, so any sequence of moves that does this must be the product of an even number of transpositions. It is therefore an even permutation.

However, the permutation required to solve the puzzle is the transposition (14 15), which is an odd permutation. Therefore no solution exists.

In effect, this proof constructs an invariant—a property of the state of the puzzle that is unchanged by any move. Define the

Symmetry

78

parity of an integer to be 0 if it is even, 1 if it is odd. Parities can be added together modulo 2: $0 + 0 = 0$, $1 + 0 = 0 + 1 = 1$, and $1 + 1 = 0$. The squares on the chessboard can also be assigned a parity: 0 for white, 1 for black. The invariant is then the parity of the number of moves *plus* that of the empty square. Any move changes each by 1, so their sum changes by $1 + 1 = 0$. The initial position has invariant 0; the required final position has invariant 1. This proves the impossibility.

It is fairly straightforward to prove that this parity sum is the *only* invariant; that is, if two positions of the puzzle have the same invariant, then there exists a sequence of moves from one to the other. So legal moves, starting from any initial state, can reach exactly *half* of the 16! possible rearrangements. Would-be solvers can potentially explore $16!/2 = 10,461,394,944,000$ arrangements, which is so large that they will always be aware that there are plenty of possibilities left. This could encourage them to think that *any* arrangement must be possible.

Rubik cube

The number of distinct arrangements of the Rubik cube, obtainable from the standard configuration, is equal to the order of the group of transformations obtained by composing rotations of the six faces. We call this the Rubik group. To compute its order, we first calculate the total number of rearrangements, ignoring the constraints of the puzzle. That is, we think about taking the puzzle apart and then reassembling it. Then we work out what fraction of those arrangements can be reached from the standard position by legal moves.

Some terminology is useful. The twenty-seven component sub-cubes are called *cubies* by aficionados. Their faces, the small coloured squares, are *facets*. There are four kinds of cubie: the centre cubie which is never visible, the centres of the faces (face cubies), those in the middle of an edge (edge cubies), and those at the corners (corner cubies). The centre and face cubies play no significant role: the centre one is fixed and the face cubies rotate but do not move. So we restrict attention to the twelve edge cubies and eight corner cubies, and assume the centre and face cubies are as in the standard configuration.

There are 8! ways to reorder the corner cubies. Each can be rotated into three different orientations. So the total number of arrangements, taking colours into account, is $3^8 8!$. Similarly the number of arrangements of the face cubies is $2^{12} 12!$. So the potential symmetry group has order:

$$3^8 8! 2^{12} 12! = 519\ 024\ 039\ 293\ 878\ 272\ 000$$

We claim that the actual symmetry group is one-twelfth as big. So its order is:

$$3^8 8! 2^{12} 11! = 43\ 252\ 003\ 274\ 489\ 856\ 000$$

The proof involves three invariants, which impose conditions on the cubies and their colours:

- *Parity on cubies*. Figure 35 (left) shows one face of the cube, with all but the central facet marked with the numbers 1–8, and the result of a clockwise quarter-turn. The corresponding permutation is

$$\begin{pmatrix} 1 & 2 & 3 & 4 & 5 & 6 & 7 & 8 \\ 7 & 8 & 1 & 2 & 3 & 4 & 5 & 6 \end{pmatrix}$$

with cycle decomposition (1753)(2864). Each 4-cycle is odd, so the product is even. All other cubies are fixed, so any quarter-turn has even parity. Therefore any element of the Rubik group has even parity as a permutation of cubies.

- *Parity on edge facets*. Figure 35 (middle) shows a similar labelling of the eight facets of the edge cubies on one layer of the cube. A quarter-turn produces the same permutation of these facets, and leaves all other edge facets fixed. So any element of the Rubik group has even parity as a permutation of edge facets.

Note that this is an extra restriction. Leaving all edge cubies fixed but flipping the facets of one has even parity on cubies, but is odd on edge facets.

- *Triality on corners*. Number the twenty-four facets of the corners so that those on two opposite faces are labelled 0, and at every corner the numbers cycle clockwise in the order 0, 1, 2, as in Figure 35 (right). Let T be the total of the numbers on any pair of opposite faces, considered modulo 3. Here the totals are 0 and 6, but these reduce to 0 modulo 3. We call T the *triality* of the

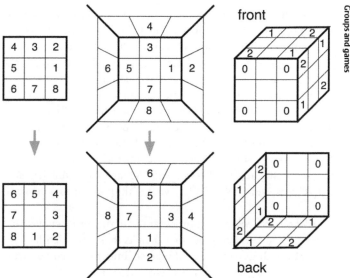

35. Invariants of the Rubik group. *Left*: Effect of a clockwise quarter-turn on cubies. *Middle*: Labelling edge facets. *Right*: Labelling corner facets

arrangement. It can be checked that any quarter-, half- or three-quarter-turn of a face leaves all faces with a total that is 0 modulo 3. So the Rubik group preserves triality, and any legal arrangement has triality 0. It is, however, easy to find illegal arrangements with triality 1 or 2: just rotate one corner cubie and leave all else fixed.

These invariants correspond to three homomorphisms from the potential symmetry group G to $\mathbf{Z}_2$, $\mathbf{Z}_2$, and $\mathbf{Z}_3$ respectively. They therefore correspond to three normal subgroups N_1, N_2, and N_3, whose orders are respectively $|G|/2$, $|G|/2$, and $|G|/3$. As already observed, in different language, N_1 and N_2 are different. The same goes for N_3 because 3 is prime to 2. Basic group theory now tells us that the intersection $N = N_1 \cap N_2 \cap N_3$ is a normal subgroup of G and $|N| = |G|/12$. (Here $12 = 2.2.3$.)

All three invariants are 0 for elements of the Rubik group, so it must be contained in N. A detailed and lengthy analysis shows that in fact the Rubik group is equal to N. The basic idea is to find enough sequences of moves to arrange almost all cubies and facets arbitrarily, and then observe that the remaining ones are determined by the three invariants. Group theory can be used to advantage in constructing such moves. For details see Tom Davis, 'Group theory via Rubik's cube' 2006 (http://www.geometer.org/rubik/index.html); Ernö Rubik, Tamás Varga, Gerazon Kéri, György Marx, and Tamás Vekerdy, *Rubik's Cubic Compendium*; and David Singmaster, *Notes on Rubik's Magic Cube* (full details in Offline Reading). We give a flavour of the group theory below.

It turns out that seven of the orientations of the corner cubies can be chosen independently, but they then fix that of the eighth corner by triality, so that only $8!3^7$ possibilities for the corners can be realized by legal moves. The twelve edge cubies can be permuted in only $12!/2$ ways by parity on cubies. Of these edge cubies, eleven can be flipped independently, but the last is then fixed by parity on edge facets. Counting these possibilities we find

$8!3^712!2^{10}$ rearrangements. This is equal to the order of N, so N is the Rubik group.

Group theory helps to solve the Rubik cube. In particular, the concept of conjugate transformations is very widely used, not always explicitly. Aficionados learn 'macros', combinations of moves that produce some specific effect. For example, the macro might flip two adjacent edge cubies while leaving everything else unchanged. By 'adjacent' I mean they are next to the same corner cubie. Now, suppose you want to flip two *non*-adjacent edge cubies while leaving everything else unchanged. Your macro doesn't work, but a conjugate does. Perform a sequence of moves s that places those two edge cubies adjacent to each other. This jumbles everything else up, but ignore that, it will all come out in the wash. Since the edge cubies are now adjacent, you can use your macro m. Finally, reverse s to get s^{-1}. All of the jumbling gets undone, except for the two edge cubies you started with—and those are flipped. How did you achieve that? Using the sequence $s^{-1}ms$, which is a conjugate of your macro m.

I'll end the discussion of Rubik's cube with a natural question, whose answer requires a detailed understanding of its symmetry group. Find the minimum number of moves that will restore any starting position to the standard one, where a move is a twist of a single face through any number of right angles. The exact minimum became known as God's number because for a long time it looked as though only an omniscient deity could work it out. But in 2010 a team of mathematicians, engineers, and computer scientists—Tomas Rokicki, Herbert Kociemba, Morley Davidson, and John Dethridge—used 350 CPU-years of computer time donated by Google to show that God's number is 20.

Sudoku

Sudoku is a combinatorial puzzle, requiring symbols to be arranged according to specific rules. The use of digits as symbols is a convenient choice, but the puzzle involves no arithmetic. Its solution involves chains of logical deductions—intelligent trial and error, eliminating incorrect choices—and can be formalized into computer algorithms. These algorithms are often used to design and check sudoku puzzles.

The history of sudoku is often traced back to Leonhard Euler in 1783. He was familiar with magic squares, in which numbers are arranged in a square grid so that all rows and columns have the same total. Euler's article 'A new type of magic square' was a variation on this theme. A typical example is:

$$
\begin{array}{ccc}
1 & 2 & 3 \\
2 & 3 & 1 \\
3 & 1 & 2
\end{array}
$$

The row and column sums are all the same, namely 6, so this is a magic square, though it violates the traditional condition of using consecutive numbers once each, and diagonal sums don't work. It is an example of a *Latin square*: an arrangement of n symbols on an $n \times n$ square grid, so that each symbol appears exactly once in every row and column. The name arises because the symbols need not be numbers, and in particular can be Latin—that is, Roman—letters.

Euler was after something more ambitious, and wrote:

> A very curious problem, which has exercised for some time the ingenuity of many people, has involved me in the following studies, which seem to open up a new field of analysis, in particular the study of combinations. The question revolves around arranging 36 officers to be drawn from 6 different ranks and also from 6 different regiments so that they are arranged in a square so that in each line (both horizontal and vertical) there are 6 officers of different ranks and different regiments.

His puzzle asks for two 6×6 Latin squares, one for each of two sets of symbols. It requires them to be *orthogonal*: each pair of symbols occurs exactly once. Euler couldn't solve his puzzle, but he did construct orthogonal $n \times n$ Latin squares for all odd n and all multiples of 4. It is easy to prove that no such squares exist for order 2. That left only $n = 6, 10, 14, 18$, and so on. Euler conjectured that no orthogonal pairs of Latin squares exist in these cases. In 1901 Gaston Tarry proved his conjecture for 6×6 squares, but in 1959 Ernest Tilden Parker constructed two orthogonal 10×10 Latin squares. In 1960 Parker, Raj Chandra Bose, and Sharadachandra Shankar Shrikhande proved that Euler's conjecture is false for all larger sizes.

A completed sudoku grid is a special type of Latin square: the 3×3 subsquares introduce extra constraints. How many different sudoku grids are there? In 2003 the answer

6,670,903,752,021,072,936,960

was posted on the USENET newsgroup *rec.puzzles,* but without a full proof. Bertram Felgenhauer and Frazer Jarvis gave a detailed calculation in 2005, with computer assistance, and relying on a few plausible but unproved assertions. The number of 9×9 Latin squares is about a million times as large. However, a sudoku grid has several types of symmetry: ways to rearrange a given grid while observing all the rules. The most obvious symmetries are the permutations of the nine symbols, the symmetric group $\mathbf{S}_9$. In addition, rows can be permuted provided the three-block structure is preserved; so can columns; and the entire grid can be reflected in its diagonal. It can be shown that the symmetry group has order $2.6^8 = 3,359,232$.

This group comes into play when we ask a basic question: how many *distinct* grids are there if we consider symmetrically related ones to be equivalent? In 2006 Jarvis and Ed Russell computed this number as:

5,472,730,538

This is not the full number divided by 3,359,232 because some grids have nontrivial symmetries.

The key to such calculations is the orbit-counting theorem, often referred to as Burnside's Lemma. Suppose that a group G acts by permutations on a set X. Given any element $x \in X$, we can apply all the elements of g to it, obtaining $g(x)$. The set of all such elements is called the *orbit* of G on X. Orbits are either the same or disjoint, so they partition X. The orbit-counting theorem states that the number of distinct orbits is

$$\frac{1}{|G|} \sum_{g \varepsilon G} |\text{Fix}_X(g)|$$

Symmetry

where $\text{Fix}_x(g)$ is the number of elements fixed by g. That is, elements x such that $g(x) = x$.

Here's a simple example of this theorem in action. There are $2^4 = 16$ ways to colour a 2×2 chessboard with black and white; Figure 36 shows them all. However, many of these colourings are equivalent under symmetries of the 2×2 square board. For example, numbers 2, 3, 5, 9 are all rotations of the same basic

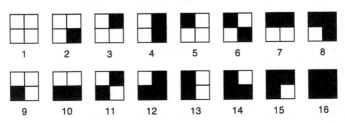

36. The sixteen ways to colour a 2×2 chessboard

Table 6. Orbit data for D_4

Identity	All sixteen patterns	16
Rotation by $\pi/2$	1, 16	2
Rotation by π	1, 6, 11, 16	4
Rotation by $3\pi/2$	1, 16	2
Reflection in horizontal axis	1, 4, 13, 16	4
Reflection in vertical axis	1, 7, 10, 16	4
Reflection in \diagonal	1, 2, 5, 6, 11, 12, 15, 16	8
Reflection in /diagonal	1, 3, 6, 8, 9, 11, 14, 16	8

pattern. Table 6 lists the eight elements of the symmetry group D_4, the patterns they fix, and how many of those there are.

The orbit-counting theorem tells us that the number of orbits is

$$\frac{1}{8}(16 + 2 + 4 + 2 + 4 + 4 + 8 + 8)$$

which is 6. In fact, the orbits are {1}, {2, 3, 5, 9}, {4, 7, 10, 13}, {6, 11}, {8, 12, 14, 15}, and {16}.

Relatively simple games and puzzles can have large symmetry groups, and can sometimes raise questions that can be hard to answer even with today's powerful machinery. They also illustrate how basic concepts of group theory, such as the parity of a permutation and the orbit-counting theorem, can help to solve combinatorial problems, or to prove that they have no solutions.

Chapter 6
Nature's patterns

Symmetries are widespread in the natural world, and they have a strong appeal to our innate sense of pattern. Figure 37 shows three instances in biology. On the left is a butterfly, *Morpho didius*. There are over eighty species in the genus *Morpho*, and they mainly inhabit South and Central America. In the middle is the eleven-armed sea star *Coscinasterias calamaria* found in the seas around southern Australia and New Zealand. It can be up to 30 cm (one foot) across. On the right is a *Nautilus* shell, shown in cross section. The nautilus is a cephalopod, and today six distinct species exist.

The butterfly has bilateral symmetry: it looks (almost) the same if it is reflected left/right about its central axis, as if in a mirror. Bilateral symmetry $\mathbf{D}_1$ is widespread in the animal kingdom.

37. Three symmetric organisms. *Left*: *Morpho* butterfly. *Middle*: **Eleven-armed sea star.** *Right*: *Nautilus*

38. Abraham Lincoln and his mirror image

Humans are an example: a person viewed in a mirror looks just like a person. In fine detail, humans are not perfectly symmetric—the face often looks subtly different when reflected; see Figure 38. For a start, Mirror-Lincoln parts his hair on the other side. Internally, there are other asymmetries in the human body: the heart is usually on the left, the intestines wind asymmetrically, and so on.

Figure 39 (left) shows an artificial butterfly created by taking the right-hand half of Figure 37 (left) and joining it to its mirror image in the grey vertical line. The resemblance to the original butterfly is striking. The sea star, suitably arranged on a flat plane, would be an almost perfectly symmetric eleven-sided regular star: Figure 39 (middle). Its symmetry group would be $\mathbf{D}_{11}$.

The symmetry of the spiral shell of *Nautilus* is more subtle. The shape, extended to infinity, is very close indeed to a logarithmic spiral, with equation $r = e^{k\theta}$ in polar coordinates, for a suitable constant k. Figure 39 (right) shows such a spiral superposed on the shell. If we translate the angle θ to $\theta + \phi$ for a fixed ϕ, the equation transforms into $r = e^{k(\theta+\phi)} = e^{k\phi}e^{k\theta}$. So the radius is multiplied by a fixed factor $e^{k\phi}$. A change of scale is called a *dilation*; in Euclidean geometry it plays the same role relative to similar triangles as rigid motions do relative to congruent ones.

39. *Left*: Right-hand side of butterfly plus its mirror image. *Middle*: Eleven-sided regular star polygon. *Right*: Logarithmic spiral superposed on *Nautilus* shell

The idealized *Nautilus* is not symmetric under rotation, and it is not symmetric under dilation either. However, it is symmetric under a suitable combination of them: rotate by ϕ and dilate by $e^{-k\phi}$. Indeed, this is a symmetry for any ϕ. So the symmetry group of the infinite logarithmic spiral is an infinite group, with one element for each real number ϕ. Two such transformations compose by adding the corresponding angles, so this group is isomorphic to the real numbers under addition.

Of course, symmetry in living creatures is never perfect. Mathematical symmetry is an idealized model. However, slightly imperfect symmetry requires explanation; it's not enough just to say 'it's asymmetric'. A typical asymmetric shape would be very different from its mirror image, not almost identical.

Bilateral symmetry in organisms

Why are many living organisms bilaterally symmetric? The full story is complicated and not fully understood, but here is a rough outline of some key issues. I have simplified the biology considerably to keep the story short.

Sexually reproducing organisms develop from a single cell, a fusion of egg and sperm. Initially this is roughly spherical. It then undergoes a repeated series of about ten to twelve divisions into 2, 4, 8, 16...cells, with the overall size staying much the same. The first few divisions destroy the spherical symmetry, distinguishing front from back (anteroposterior axis), top from bottom (dorsoventral axis), and left from right. In subsequent development the first two symmetries are quickly lost too, but the embryo tends to retain left–right symmetry until the organism has become fairly complex.

Development is a combination of the natural 'free-running' chemistry and mechanics of the cell with genetic 'instructions' that keep the developmental programme on track. It might seem that free-running dynamics automatically preserves left–right symmetry, but differences between the left and right sides can easily set in, so some genetic regulation is needed to maintain symmetry. Mirror-symmetric body plans evolved very early, and evolution may have selected for bilateral symmetry because this makes it simpler to control movement (imagine walking with one short leg and one long one) and because the same developmental plan can, in effect, be used twice.

The internal structure is often forced to become asymmetric for geometric or mechanical reasons. The human gut is too long to fit inside the body cavity without being folded, and no symmetric method of folding can fit it in. But there is good evidence that genes are involved as well. A number of biological molecules have been found which relay asymmetric signals. In 1998 it was discovered that the gene *Pitx2* is expressed (activated) in the left heart and gut of embryos of the mouse, chick, and *Xenopus* (a frog). Failure to express this gene correctly causes misplaced organs. In the same year it was discovered that if the protein Vg1, a growth factor known to be associated with left–right asymmetry is injected into particular cells on the right side of a *Xenopus* embryo, where this protein does not normally occur, the entire structure of the internal organs flipped to a mirror image of the usual form. Further experiments led to the idea that Vg1 is a very early step in the developmental pathway that sets up the left–right axis: whichever side gets Vg1 becomes the 'left' side in terms of normal development.

It has also been suggested that bilateral symmetry plays a role in sexual selection, an evolutionary phenomenon in which female preferences interact with male features (sometimes the other way round) to create an evolutionary 'arms race' that drives the male to develop exaggerated body forms that without this selective pressure would reduce the prospects of surviving to breed. The enormous tail of the peacock is a standard example. These preferences can be arbitrary, but any that are associated with 'good genes' will also reinforce biological fitness. Since symmetric development has a genetic component, external symmetry can function as a test for good genes. So it is natural for each sex to prefer symmetric features in the other. Experiments showed that female swallows were less attracted to males with asymmetric tails, and the same went for the wings of Japanese scorpion flies. It is often stated that movie stars have unusually symmetric faces, but this whole area is controversial because even when symmetry can be associated with preference, it is extremely difficult to establish the reasons for the association.

A great deal is known about the role of various genes in the symmetric body plans of vertebrates, echinoderms (the fivefold symmetry of starfish, for example), and flowers. In 1999 it was found that a mutation in the plant *Linaria vulgaris* can change the symmetry of the flower from bilateral to radial. The mutation affects a gene called *Lcyc*, and 'switches it off' in the mutant. The causes of symmetry in living creatures are complex and subtle.

Animal gaits

The symmetry of a living creature affects not just its shape, but how it moves. This phenomenon is particularly striking in its most familiar instance: the movements of quadrupeds, four-legged animals. Horses walk at low speeds, trot at intermediate ones, and gallop at high speeds. Many insert a fourth pattern of movement, the canter, between trot and gallop. Camels and giraffes employ another pattern, the pace. Many small animals, such as rabbits and squirrels, bound. Dogs walk, trot, and bound; cats walk and bound. Pigs walk and trot; see Figure 40. Throughout the animal kingdom, quadrupeds make use of a small, standard list of patterns of movement, known as *gaits*.

Gait analysis goes back at least to Aristotle, who argued that a trotting horse can never be completely off the ground. The subject began in earnest when Eadweard Muybridge started using arrays of still cameras to take series of photographs of humans and animals in motion. Only then was it possible to see exactly what the animal was doing. In particular, it turned out that a trotting horse can be completely off the ground during some phases of its motion. This settled a rather expensive bet in favour of Leland Stanford, former governor of California.

Gaits are periodic cycles of movement, idealized from actual animal motion, which can stop, start, and change according to decisions made by the animal. The ideal gait repeats the same *gait cycle* over and over again. If two legs follow the same cycle but one has a time delay relative to the other, then the difference in timing is called a phase shift. Here we measure such a shift using the corresponding fraction of the period.

Like all periodic motions, gaits have time-translation symmetry: change phase by any integer number of cycles. There is also a spatial symmetry, the bilateral symmetry of the animal. However, the timing patterns of gaits suggest considering another kind of symmetry, which applies to the patterns but not the animal as such: permuting the legs. For example, in the bound, both front legs hit the ground together, then both back legs, and there is a symmetry that swaps front and back, combined with a half-cycle phase shift; see Figure 41. This is not a symmetry of the animal, but it is clearly present in several gaits, and is crucial to one method for modelling and predicting gait patterns.

Gait analysts have long distinguished symmetric gaits, such as the walk, pace, and bound, from asymmetric ones, such as the canter

40. Trotting sow (Muybridge)

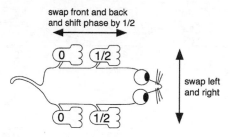

swap front and back
and shift phase by 1/2

swap left
and right

41. Spatio-temporal symmetries of the bound

and gallop. Permutational symmetries of the legs refine this classification and link the patterns to a structure in the animal's nervous system known as a central pattern generator, which is thought to control the basic rhythms of the motion. The timings for some of the common gaits can be summarized using the fractions of the gait cycle at which the four legs first hit the ground; see Figure 42. Here we employ the convention that the cycle starts when the left rear leg hits the ground, which is convenient mathematically.

Here the fractions 1/4, 1/2, 3/4, occurring in the symmetric gaits, are more or less exact and do not vary from one animal to another. The fractions 1/10, 6/10, 9/10, occurring in the asymmetric gaits, are more variable, and can change depending on the animal and the speed with which it moves.

The permutational symmetries that fix these gaits, when combined with an appropriate phase shift, can be described informally as follows:

- In the walk, legs cycle in the order LF → RF → LR → RR with phase shifts 1/4 between each successive leg.
- In the trot, corresponding diagonal pairs of legs are synchronized. Swapping front and back, or left and right, induces a phase shift 1/2.
- In the bound, corresponding left and right legs are synchronized. Swapping front and back induces a phase shift 1/2.

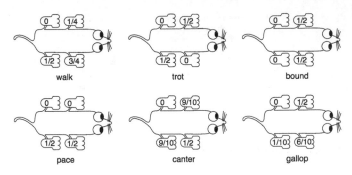

42. Spatio-temporal symmetries of some standard gaits

- In the pace, corresponding front and back legs are synchronized. Swapping left and right induces a phase shift 1/2.
- In the canter, there is a 1/2 phase shift on one diagonal pair of legs, and the other pair is synchronized.
- In the gallop, there is a 1/2 phase shift from front to back. (Left and right are almost synchronized, but not exactly.) More precisely, this gait is a transverse gallop, found, for example, in horses. The cheetah uses a rotary gallop, in which the phases of the front legs are interchanged.

These patterns are very similar to those observed in closed rings of identical oscillators. For example, if four oscillators numbered 0, 1, 2, 3 are connected successively in a ring, with each influencing the next (but not the reverse) then the main natural patterns of periodic oscillation ('primary' oscillations) are:

0	1	2	3
0	0	0	0
0	1/4	1/2	3/4
0	3/4	1/2	1/4
0	1/2	0	1/2

The second pattern resembles the walk, if oscillators are assigned to legs in a suitable way. With the same assignment, the third pattern is a backwards walk. The fourth pattern resembles the bound, pace, or trot, depending on how the oscillators are assigned to legs.

The table for the four-oscillator ring includes one pattern of phase shifts that has not yet been mentioned in connection with gaits: the first entry, with all four legs synchronized. This gait does occur in some animals, such as the gazelle, and it is called a pronk (or stot). The whole animal jumps, with all four legs leaving the ground simultaneously. This gait is thought to have evolved to confuse predators, but that suggestion is no more than speculation.

There is a plausible reason to suppose that the central pattern generator for gait patterns must possess cyclic group symmetry of this general kind, in order to generate the observed patterns in a robust manner: see M. Golubitsky, D. Romano, and Y. Wang, 'Network periodic solutions: patterns of phase-shift synchrony', *Nonlinearity* **25** (2012) 1045–74. For reasons too extensive to go into here, the central pattern generator architecture that most closely models observations consists, in a schematic description, of two rings, each composed of four 'units' of nerve cells, connected left–right an a mirror-symmetric manner. Each ring controls the basic timing of the legs on its side of the animal, but two units control the muscles of the back legs and the other two control the front legs. The units assigned to a given leg are not adjacent in the ring, but are spaced alternately. This architecture predicts all of the common gait patterns—the gallop and the canter are 'mode interactions' in which two distinct patterns compete—and, crucially, does not predict the innumerable other patterns that could be envisaged. The literature on gaits is huge, including detailed models of the mechanics of locomotion. The symmetry analysis is just one small part of a complex and fascinating area.

Sand dunes

Nature's tendency to create symmetric objects is especially striking in the flow of sand as winds blow over a desert. There's not a great deal of structure in a desert, and winds tend either to blow fairly consistently in one main direction, the 'prevailing wind', or to vary all over the place. Neither feature sounds like an ingredient that might lead to symmetry, but sand dunes exhibit striking patterns, a sign that symmetries may lurk somewhere nearby.

Geologists classify dunes into six main types: longitudinal, transverse, barchanoid, barchan, parabolic, and star; see Figure 43. Reality is less symmetric than idealized mathematical models, so all of the symmetries that I'll mention are approximate ones. In controlled experiments and computer simulations, where the

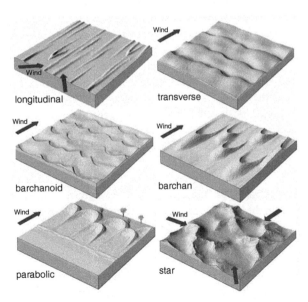

43. Sand dune shapes

'desert' is perfectly flat and uniform and the wind blows in a regular fashion, the symmetries are closer to the ideal.

Longitudinal and transverse dunes occur in equally spaced parallel rows, when there is a strong prevailing wind in a fixed direction. In effect, they are stripes in the sand. Longitudinal dunes are aligned with the wind direction. Transverse dunes are aligned at right angles to the wind direction. Barchanoid dunes are like transverse ones, but they have scalloped edges, as if the striped pattern is starting to break up.

Barchan dunes are what happen when the stripes do break up. Each dune is a crescent-shaped mound of sand, with the arms of the crescent pointing in the direction of the wind. Barchans often form a 'swarm' of nearby dunes, and in models these are often the same shape and size, and regularly spaced in a lattice. In reality, they are irregularly spaced and can have different sizes. Sand is blown up the front of the dune and falls over the top to the far side; at the tips of the crescents the sand also flows round the side. As a result, the entire dune moves slowly downwind, retaining the same shape. In Egypt, entire villages can disappear beneath an advancing barchan dune, only to re-emerge decades later when the dune moves on.

Parabolic dunes are superficially like barchans, but oriented in the opposite direction: the tips of the crescent point into the wind. They tend to form on beaches, where vegetation covers the sand. Star dunes are isolated spikily ridged hills, and again often occur in swarms. They are found when the wind direction is highly irregular, and they form starlike shapes with three or four pointed arms.

Symmetries don't just let us organize these patterns. They help us to understand how they arise; see Figure 44. Sand dunes are typical of many pattern-forming systems in mathematical physics, and they exemplify a general, powerful way of thinking about these systems. The key idea is known as *symmetry breaking*. At

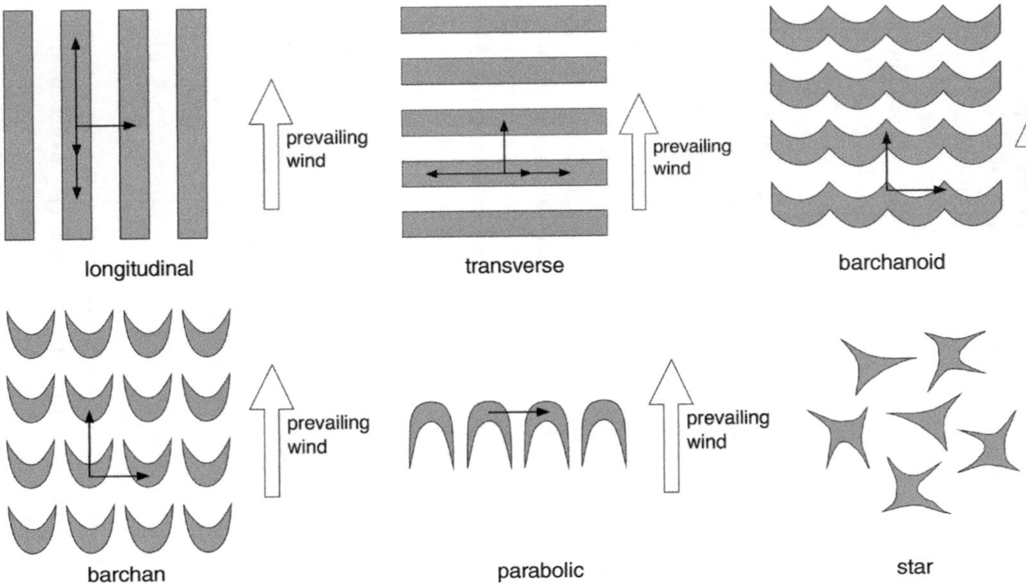

Symmetry

longitudinal

transverse

barchanoid

prevailing wind

prevailing wind

prevailing wind

barchan

parabolic

star

prevailing wind

prevailing wind

44. **Symmetries of dune patterns**

first sight it seems to violate Curie's principle (see Chapter 2), because the observed state has less symmetry than its cause does. However, a tiny asymmetric perturbation is required to create this state, so technically Curie is correct.

First, imagine a very regular model desert in which the wind blows with constant speed and constant direction over an infinite plane of initially flat sand. The symmetries of the system comprise all rigid motions of the plane that fix the wind direction. These are all translations of the plane, together with left–right reflection in any line parallel to the wind.

If the profile of the sand has full symmetry, then the sand will be the same depth everywhere, because any point can be translated to any other. So we get a uniform flat desert. However, this state can become unstable, and intuitively we expect this if the wind becomes strong enough—while remaining uniform—to disturb sand grains. Tiny random effects will cause some grains to move, while others remain in place. Tiny dips and bumps begin to appear, and these affect the flow of the wind nearby. Vortices trail off from the sides of the bumps, and local wind speeds can increase. These effects can amplify through feedback, and the symmetry breaks.

What does it break *to*? The plane has independent translational symmetries in two directions: parallel to the wind, and at right angles to it. If the translational symmetry at right angles breaks, the most symmetric possibility is that instead of the pattern being fixed by all translations, it is fixed by a subgroup: translation through integer multiples of some fixed length. The result is parallel waves, separated by the length in question. The waves are invariant under all translations parallel to the wind, so they look like an array of parallel stripes, running in the direction of the wind. These are longitudinal dunes.

If the translational symmetry along the direction of the wind breaks, much the same happens, but now the waves form at right angles to the wind. So we get transverse dunes.

Barchanoid dunes form when a second symmetry breaks: the group of all translations at right angles to the wind. Again, this becomes a discrete subgroup, creating the rippled pattern along the ridges. The ripples are equally spaced, and they are all the same shape. The group of translational symmetries is lattice: its generator moves the entire ridge either one step forwards or one step sideways. Each ripple is also bilaterally symmetric in all mirror lines parallel to the wind and passing through either the apex of each ripple or the point halfway between two apexes. So some of the reflectional symmetries break, but some do not.

Individual barchan dunes also have this kind of mirror-image symmetry, and so do theoretical arrays of barchan dunes. Detailed models of the flow of air and sand explain the crescent shape, which is caused by a large vortex that separates from the overall flow across the dune.

Parabolic dunes break translation along the wind direction altogether: they are pinned in place by the edge of the beach. They are symmetric under a discrete group of sideways translations and reflections similar to those found for barchans.

Star dunes form when there is *not* a prevailing wind direction. It's probably best to say that they have lost all symmetry, but they have traces of rotational symmetry—their starlike shape—which may correspond to the rotational symmetry of the *average* wind direction: equally likely to blow in any direction whatever.

When we think about the symmetries of the patterns, and how each relates to the others, we start to see a degree of order in what might otherwise seem a very disordered catalogue. The group-theoretic analysis of symmetries, and how they break, reveals a deeper structure. Ironically, Curie's principle applies only in a world that is even more idealized than the mathematical model of symmetric equations plus small random perturbations: namely, a world that lacks the random perturbations. It tells us that any explanation of the patterns must involve something that breaks the symmetry, but it doesn't explain any of the patterns that then appear.

Galaxies

Galaxies have beautiful shapes, but are they symmetric? I'm going to argue that the answer is 'yes', but the reasoning depends on modelling assumptions and which kinds of symmetry are being considered.

The most dramatic feature of a galaxy is its spiral form. This is often close to a logarithmic spiral—for example, the spiral arms of our own galaxy, the Milky Way, are roughly of this form. When discussing the *Nautilus* shell we saw that the logarithmic spiral has a continuous family of symmetries: dilate by some amount and rotate through a corresponding angle. Strictly speaking, this symmetry applied only to the complete infinite spiral. Real galaxies are of finite extent, and a finite spiral cannot have dilation-plus-rotation symmetry. However, it is reasonable and commonplace to model finite patterns as portions of ideal infinite ones, so this objection carries little weight. We've just used this kind of model for sand dunes, in fact. A more significant objection is that the spiral arms do not extend far enough to confirm that the spiral really is close to logarithmic.

A glance at pictures of galaxies shows that many of them have a remarkably close approximation to symmetry under rotation through 180°. Figure 45 shows two examples, the Pinwheel Galaxy and NGC 1300. The former is a spiral, the latter a barred

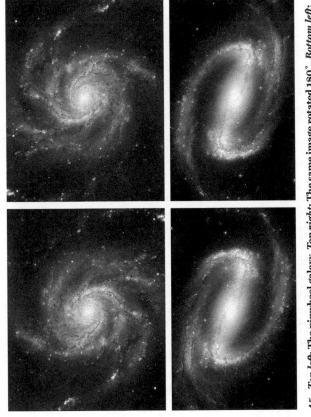

45. *Top left*: The pinwheel galaxy. *Top right*: The same image rotated 180°. *Bottom left*: The barred-spiral galaxy NGC 1300. *Bottom right*: The same image rotated 180°

spiral. The pictures show an image of each galaxy next to the same image rotated 180°. At first glance, it is hard to tell the difference.

According to many of the current mathematical models of galaxy dynamics, the arms of a spiral or barred spiral galaxy are probably rotating waves, which retain the same shape as time passes, but rotate about the galaxy's centre. (It has also been suggested that barred spirals may be created by chaotic dynamics: see Panos A. Patsis, 'Structures out of chaos in barred-spiral galaxies', *International Journal of Bifurcation and Chaos* D-11-00008.) The waves are thought to be density waves, so the densest regions do not always contain the same stars. A sound wave is a density wave: as sound passes through the air, some parts become compressed, and the compressed region travels like a wave. However, the molecules of air do not travel with the compression wave; they remain close to their original position. Whether the spiral arms are waves of stars or waves of density, rotating waves have a continuous family of space-time symmetries: wait for a period of time and rotate through a suitable angle. So in fact galaxies are highly symmetric, and the symmetry constrains their form.

Most galaxies with (approximate) rotational symmetry are fixed by 180° rotation, but a few seem to have higher-order rotational symmetries. A three-armed spiral could be symmetric under a 120° rotation, for example. This seems to be very rare in real galaxies, but it occurs in some simulations and is observed in the galaxy NGC 7137. The Milky Way's spiral arms have approximate 90° rotational symmetry, but the existence of a central bar reduces this to 180° rotation only.

Snowflakes

In 1611 Johannes Kepler, an inveterate pattern seeker, gave his sponsor Matthew Wacker a New Year present: a small book that he had written with the title *De Nive Sexangula* ('On the

46. Snowflakes photographed by Vermont farmer Wilson Bentley, published in *Monthly Weather Review* 1902

six-cornered snowflake'). Kepler's main target was the notorious six-sided symmetry of snowflakes, made all the more baffling by the enormous variety of shapes that occur; see Figure 46. Notice that the bottom-right image has threefold symmetry, not sixfold, showing that other shapes are also possible.

Kepler deduced, on the basis of thought experiments and known facts, that the 'formative principle' for a snowflake must be related to closely packed spheres—much as a number of pennies on a table naturally pack into a honeycomb pattern. The current explanation is along those lines: the crystal lattice of the relevant form of ice consists of slightly bumpy layers whose main symmetry is hexagonal. This creates a six-sided 'seed' upon which the snowflake grows. The exact shape is affected by the temperature and humidity of the storm cloud, which vary chaotically, but because the flake is very small compared to the scales on which these quantities vary, very similar conditions occur at all six corners. So the sixfold (that is, $\mathbf{D}_6$) symmetry is maintained to a good approximation. However, instabilities can break this symmetry, and other physical processes come into play under different meteorological conditions.

Other patterns

Many other forms and patterns in the natural world are evidence for the symmetry of the processes that generate them. The Earth is roughly spherical because it condensed out of a disc of gas surrounding the nascent Sun. The natural minimum-energy configuration for a ball of molten rock is a sphere, and this is related to the symmetry of the condensation process about the centre of mass. On a closer level of detail, the Earth is flattened at the poles because it was rotating on its axis while still molten. The symmetry broke from spherical to circular, with a top–bottom reflection, to produce an ellipsoid of rotation.

In 1956 Alan Turing, famous for his wartime work at Bletchley Park on the enigma code, wrote a paper suggesting a mathematical model for the formation of animal markings, such as the spots on a leopard and the stripes on a tiger. His idea was that some diffusing and reacting system of chemicals, which he called morphogens, laid down an invisible pre-pattern in the embryo; later this was turned into a visible pattern by the production of pigment proteins according to the pre-pattern. A great deal of work has been done on these reaction-diffusion equations. They turn out to be too simple to capture biological reality, but they have been extended to more realistic models. Most natural animal markings can be produced by equations of this kind, and Hans Meinhardt has made an extensive study of the markings on seashells from this point of view in *The Algorithmic Beauty of Sea Shells*. More work needs to be done to include genetic effects and make the models even more realistic, but some steps in that direction have already been taken.

Laboratory experiments have revealed a remarkable pair of patterns in reaction-diffusion equations, arising in the Belousov-Zhabotinskii reaction, named after its discoverers

Boris Belousov (in the 1950s) and Anatol Zhabotinskii (a rediscovery in 1961). If three particular chemicals are mixed together in a shallow dish, along with a fourth that changes from blue to red according to whether the reaction is oxidizing or reducing, the liquid turns blue, then a uniform orange-red. After a few minutes, however, random spots of blue appear, and expand. When they become large enough, red spots appear in their centres, and soon the dish contains several slowly expanding 'target patterns' with successive rings of blue and red. If disturbed, these rings can break, and curl up into rotating spirals, which also slowly grow. The B-Z reaction has been widely studied, and a few papers have applied symmetry-breaking methods. These predict that three symmetry types of time-periodic pattern are especially natural: one with circular symmetry, one a rotating wave, and one with reflectional symmetry. All can be stable, but the rotating wave and the state with reflectional symmetry should not both be stable simultaneously. Further analysis has related the state with circular symmetry to target patterns, and the rotating wave to spirals.

Similar waves of electrical activity occur in pacemaker signals sent to the muscles of the heart to control how it beats. Here target patterns are normal, and spirals can be fatal. So we all carry around inside ourselves a system whose symmetry-breaking dynamics are, literally, of vital importance.

Chapter 7
Nature's laws

Albert Einstein remarked that the most surprising thing about Nature is that it is comprehensible. He meant that the underlying laws are simple enough for the human mind to understand. How Nature behaves is a consequence of these laws, and simple laws can generate extremely complex behaviour. For example, the movement of the planets of the solar system is governed by the laws of gravity and motion. These laws (either in Newton's version or in Einstein's) are simple, but the solar system is not.

The term 'law' here has a misleading air of finality. All scientific laws are provisional: approximations that are valid to a high degree of accuracy and are used until something better comes along.

One of the most intriguing features of the laws of Nature, as we understand them, is that they are symmetric. As we saw in the previous chapter, symmetry of the equations (laws) need not imply symmetry of the behaviour (solutions). In general, the laws of Nature are more symmetric than Nature itself, but the symmetries of the laws can break. The patterns exhibited by the behaviour provide clues to the symmetries that are being broken.

Physicists in particular have found this observation to be of vital importance when trying to find new laws of Nature.

* * *

One of the fundamental theorems in this area is Noether's Theorem, proved by Emmy Noether in 1918. A Hamiltonian system is a general form of equation for mechanics without frictional forces. The theorem states that whenever a Hamiltonian system has a continuous symmetry, there is an associated conserved quantity. 'Conserved' means that this quantity remains unchanged as the system moves.

For example, energy is a conserved quantity. The corresponding continuous symmetry—that is, group of symmetries parametrized by a continuous variable—is time translation. The laws of Nature are the same at all times: if you translate time from t to $t + \theta$ the laws don't look any different. Following through the nuts and bolts of Noether's proof, the corresponding conserved quantity is energy. Translation in space (the laws are the same everywhere) corresponds to conservation of momentum. Rotations are another source of continuous symmetries; here the conserved quantity is angular momentum about the axis of rotation. The profound conservation laws discovered by the classical mechanicians—Newton, Euler, Lagrange—are all consequences of symmetry.

* * *

The standard setting for the study of continuous symmetries is Lie theory, named after the Norwegian mathematician Sophus Lie. The resulting structures are Lie groups, with which are associated Lie algebras. To motivate the main ideas, we consider one example, the *special orthogonal group* **SO**(3). This consists of all rotations in three-dimensional space. A rotation is specified by its axis, which remains fixed, and an angle: how big the rotation is. These variables are continuous: they can take any real value. So this group has a natural topological structure as well as its group structure. Moreover, the two are closely linked: if two pairs of group elements are very close together, so are their products.

That is, the group operations are continuous maps. Indeed, more is true: we can apply the operations of calculus, in particular taking the derivative. The group operations turn out to be differentiable.

More strongly, the group has a geometric structure analogous to that of a smooth surface but with more dimensions. To find its dimension, observe that it takes two numbers to specify the axis of rotation (say the longitude and latitude of the point in which the axis meets the northern hemisphere of the unit sphere) and one further number to specify the angle. So without doing any serious calculations, we know that **SO**(3) is a three-dimensional space.

Algebraically, **SO**(3) can be defined as the group of all 3×3 orthogonal matrices of determinant 1. A matrix M is orthogonal if $MM^T = I$ where I is the identity matrix and T indicates the transpose. There is an important connection with another type of matrix. The exponential of any matrix M can be defined using the convergent series

$$\exp M = I + \frac{1}{2!}M^2 + \frac{1}{3!}M^3 + \ldots + \frac{1}{n!}M^n + \ldots$$

and a simple calculation shows that every matrix in **SO**(3) is the exponential of a skew-symmetric matrix, for which $M^T = -M$, and conversely.

The product of two orthogonal matrices is always orthogonal, but the product of two skew-symmetric matrices need not be skew-symmetric. However, the *commutator*

$$[L, M] = LM - ML$$

of two skew-symmetric matrices is always skew-symmetric. A vector space of matrices that is closed under the commutator is called a *Lie algebra*. So we have associated a Lie algebra with the special orthogonal group, and the exponential map sends the Lie algebra to the group.

More generally, a *Lie group* is any group that also has a particular type of geometric structure, with respect to which the group operations (product, inverse) are smooth maps. Every Lie group has an associated real Lie algebra, which describes the local structure of the group near the identity element. This in turn determines a complex Lie algebra. Using complex Lie algebras, it is possible to classify—that is, determine the structure of—some important types of Lie group. The first step is to classify the simple complex Lie algebras, which are complex Lie algebras L that do not contain a subalgebra K (other than 0 or L) such that $[L,K] \subseteq K$. Such a subalgebra is called an ideal, and this property is the analogue for a Lie algebra of a normal subgroup.

In 1890 Wilhelm Killing obtained a complete classification of all simple complex Lie algebras, subject to a few errors and omissions that were soon corrected. This classification is now presented in terms of graphs known as *Dynkin diagrams*, which specify certain geometric structures called *root systems*. Every simple complex Lie algebra has a root system, and this completely determines its structure. Figure 47 shows the Dynkin diagrams. There are four infinite families, denoted A_n ($n \geq 1$), B_n ($n \geq 2$), C_n ($n \geq 3$), and D_n ($n \geq 4$). In addition, there are five exceptional diagrams, denoted G_2, F_4, E_6, E_7, and E_8. The dimensions of these algebras (as vector spaces over $\mathbf{C}$) are listed in Table 7.

The four infinite families that occur in the classification theorem can be realized as Lie algebras of matrices under the commutator operation. The type A_n algebra is the special linear Lie algebra $\mathbf{sl}_{n+1}(\mathbf{C})$, consisting of all $(n+1) \times (n+1)$ complex matrices whose trace (sum of diagonal terms) is zero. The type B_n algebra consists of skew-symmetric $(2n+1) \times (2n+1)$ complex matrices, denoted $\mathbf{so}_{2n+1}(\mathbf{C})$. The type D_n algebra consists of skew-symmetric

Table 7. The classification of the simple complex Lie algebras

Lie algebra	Dimension
A_n	$n(n+2)$
B_n	$n(2n+1)$
C_n	$n(2n+1)$
D_n	$n(2n-1)$
G_2	14
F_4	52
E_6	78
E_7	133
E_8	248

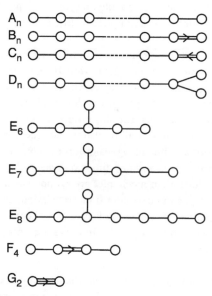

47. Dynkin diagrams

113

$2n{\times}2n$ complex matrices, denoted $\mathbf{so}_{2n}(\mathbf{C})$. And the type $\mathbf{C}_n$ algebra consists of *symplectic* $2n{\times}2n$ complex matrices, denoted $\mathbf{sp}_{2n}(\mathbf{C})$. These are the matrices that can be written in block form as:

$$\begin{bmatrix} X & Y \\ Z & -X^{\mathrm{T}} \end{bmatrix}$$

where X, Y, Z are $n{\times}n$ matrices and Y and Z are symmetric.

The complex simple Lie algebras are fundamental to the classification of simple Lie groups, but the passage from the real numbers to the complex numbers introduces some complications because the geometric structure of a Lie group is defined in terms of real coordinates. Each simple Lie algebra has a variety of 'real forms', and these correspond to different groups. Moreover, for each real form, there is still some freedom in the choice of group: groups that are isomorphic modulo their centres have the same Lie algebra. Nonetheless, a complete picture can be derived.

* * *

Lie groups are not always simple. A familiar example, one that we have been studying from time to time throughout this book, without giving it a technical name, is the *Euclidean group* $\mathbf{E}(2)$ of all rigid motions of the plane. This has a subgroup $\mathbf{R}^2$ that comprises all translations, and this group turns out to be normal. $\mathbf{E}(2)$ also contains all rotations and reflections, and it is three dimensional. The analogous group $\mathbf{E}(n)$ in dimension n has similar properties, and has dimension $n(n + 1)/2$. The equations of Newtonian mechanics are symmetric under the Euclidean group, and also under time translation, and Noether's Theorem explains the existence of the classical conserved quantities as consequences of continuous subgroups, as described above.

Another important group in classical (that is, non-relativistic) mechanics is the Galilean group, which is used to relate two different coordinate systems (frames of reference) that are moving at uniform velocity relative to each other. Now we need transformations that correspond to uniform motion, in addition to those in the Euclidean group.

From the modern point of view, the most influential symmetries in classical mechanics are those that relate to its reformulation by William Rowan Hamilton, in terms of a single function. We call it the Hamiltonian of the system. It can be interpreted as its energy, expressed as a function of position and momentum coordinates. The appropriate transformations turn out to be symplectic. Most advanced research in classical mechanics is now done in the framework of symplectic geometry.

Another Lie group that is very similar to the Euclidean group arises in special relativity. Here, the usual squared-distance function

$$d^2 = x^2 + y^2 + z^2$$

in three-dimensional space is replaced by the *interval* between events in space-time:

$$d^2 = x^2 + y^2 + z^2 - c^2 t^2$$

where t is time.

The scaling factor c^2 merely changes the units of time measurement, but the minus sign in front of it changes the mathematics and physics dramatically. The group of transformations of space-time that fixes the origin and leaves the interval invariant is called the *Lorentz group* after the physicist Hendrik Lorentz. The Lorentz group specifies how relative motion works in relativity, and is responsible for the theory's counterintuitive features in which objects shrink, time slows down, and mass increases, as a body nears the speed of light.

Just over a century ago, most scientists did not believe that matter was made of atoms. As experimental and theoretical support grew, atomic theory became first respectable, then orthodox. Atoms, at first thought to be indivisible—which is what the word means, in Greek—turned out to be made from three kinds of particle: electrons, protons, and neutrons. How many of each an atom possessed determined its chemical properties and explained Dmitri Mendeleev's periodic table of the elements. But soon other particles joined the game: neutrinos, which rarely interact with other particles and can travel through the Earth without noticing it's there; positrons, made of antimatter, the opposite of an electron; and many more. Soon the zoo of allegedly 'elementary' particles contained more particles than the periodic table contained elements.

At the same time, it became clear that there are four basic types of force in Nature: gravity, electromagnetic, weak nuclear, and strong nuclear. Forces are 'carried' by particles, and particles are associated with quantum fields. Fields pervade the whole of space, and change over time. Particles are tiny localized clumps of field. Fields are seething masses of particles. A field is like an ocean, a particle is like a solitary wave. A photon, for instance, is the particle associated with the electromagnetic field. Waves and particles are inseparable: you can't have one without the other.

As this picture slowly assembled, step by step, the vital role played by symmetry became increasingly prominent. Symmetries organize quantum fields, and therefore the particles associated with them. Out of this activity emerged the best theory we have of the truly fundamental particles; see Figure 48. It is called the standard model. The particles are classified into four types: fermions and bosons (which have different statistical properties), quarks and leptons. Electrons are still fundamental, but protons and neutrons are not: they are composed of quarks of six different kinds. There are three types of neutrino, and the electron is accompanied by two other particles, the muon and tauon. The

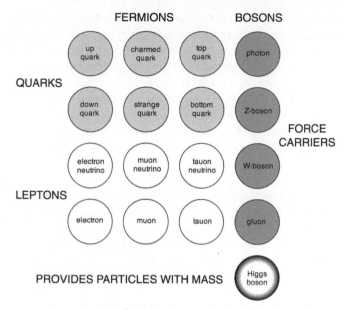

FERMIONS   BOSONS

QUARKS

| up quark | charmed quark | top quark | photon |
| down quark | strange quark | bottom quark | Z-boson |

FORCE CARRIERS

LEPTONS

| electron neutrino | muon neutrino | tauon neutrino | W-boson |
| electron | muon | tauon | gluon |

PROVIDES PARTICLES WITH MASS   Higgs boson

48. Particles of the standard model

photon is the carrier for the electromagnetic force; the Z- and W-bosons carry the weak nuclear force; the gluon carries the strong nuclear force.

As described, the theory predicts that all particles have zero mass, and this is not consistent with observations. The final piece in the jigsaw is the Higgs boson, which endows particles with masses. The field corresponding to the Higgs boson differs from all others in that it is nonzero in a vacuum. As a particle moves through the Higgs field, its interaction with the field endows it with behaviour that we interpret as mass. In 2012 a new particle consistent with the theoretical Higgs boson was detected by the Large Hadron Collider. Further observations will be required to decide whether it corresponds exactly to the predicted particle, or is some variant that might lead to new physics.

Symmetries are crucial to the classification of particles because the possible states of a quantum system are to some extent determined by the symmetries of the underlying equations. Specifically, what matters is how the group of symmetries acts on the space of quantum wave functions. The 'pure states' of the system—states that can be detected when observations are made—correspond to special solutions of the equations, called eigenfunctions, which can be worked out from the symmetry group. The mathematics is sophisticated, but the story can be understood in general terms, as follows.

A useful analogy is Fourier analysis, which represents any 2π-periodic function as a linear combination of sines and cosines of integer multiples of the variable. Passing to the complex numbers, any 2π-periodic function is represented by an infinite series of exponentials e^{nix} with complex coefficients. The relevant symmetry group here consists of all translations of x modulo 2π, which physically represent phase shifts of the periodic function. The resulting group $\mathbf{R}/2\pi\mathbf{Z}$ is isomorphic to the circle group $\mathbf{SO}(2)$, so the whole set-up is symmetric under the phase-shift action of $\mathbf{SO}(2)$ on the vector space of all 2π-periodic functions. Fourier analysis originated in work on the heat equation and the wave equation in mathematical physics, and these equations have $\mathbf{SO}(2)$ symmetry, realized as phase shifts on periodic solutions. The solutions e^{nix}, for specific n, are special solutions; in the context of the wave equation these functions—rather, their real parts—are especially familiar as normal modes of vibration. In music, the vibrating object is a string, and the normal modes are the fundamental note and its harmonics.

For a deeper interpretation of the mathematics, we consider how $\mathbf{SO}(2)$ acts on the space of periodic functions. This is a real vector space, of infinite dimension. The normal modes span subspaces, which are two-dimensional except for the zero mode when the subspace is one-dimensional. A (real) basis for this space consists of the functions $\cos nx$ and $\sin nx$, except when $n = 0$, in which case the sine term is omitted because it is zero, and the cosine is constant. Each such subspace is invariant under the symmetry group—that is, a phase shift applied to a normal-mode wave is a normal-mode wave. This is most easily verified in complex coordinates, because $e^{ni(x+\phi)} = e^{ni\phi}e^{nix}$, and $e^{ni\phi}$ is just a complex constant. In real coordinates, both $\cos x + \phi$ and $\sin x + \phi$ are linear combinations of $\cos x$ and $\sin x$.

Geometrically, the action of $\theta \in \mathbf{SO}(2)$ on the subspace spanned by e^{nix} is rotation through an angle $n\theta$. So each subspace provides a *representation* of $\mathbf{SO}(2)$, that is, a group of linear transformations that is isomorphic to it, or more generally a homomorphic image of it. The linear transformations correspond to matrices, and the representation is *irreducible* if no proper nonzero subspace is invariant under (mapped to itself by) every such matrix. So what Fourier analysis does, from the point of view of symmetry, is to decompose the representation of $\mathbf{SO}(2)$ on the space of 2π-periodic functions into irreducible representations. These representations are all different, thanks to the integer n.

This set-up can be generalized, with $\mathbf{SO}(2)$ replaced by any compact Lie group. A basic theorem in representation theory states that any representation of such a group can be decomposed into irreducible representations. Notice that the normal mode e^{nix} is an eigenvector for all of the matrices given by the group, again because $e^{ni(x+\phi)} = e^{ni\phi}e^{nix}$, and $e^{ni\phi}$ is a constant.

Quantum mechanics is similar, but the wave equation is replaced by Schrödinger's equation or equations for quantum fields. Complex numbers are built into the formalism from the start. The analogues of normal modes are eigenfunctions. So every solution of the equation, that is, every quantum state for the system being modelled, is a linear combination—a superposition—of eigenfunctions. Experiment and theory suggest that superposed states should not be observable as such; only individual eigenfunctions can be observed. More precisely, observing a superposition is delicate and only possible in unusual circumstances; until recently it was believed to be impossible. Associated with this suggestion is the Copenhagen interpretation, in which any observation somehow 'collapses' the state to an eigenfunction. This proposal led to quasi-philosophical ideas such as Schrödinger's cat and the many-worlds interpretation of quantum mechanics. All we need here, however, is the underlying mathematics, which tells us that observable states correspond to irreducible representations of the symmetry group of the equation. In particle physics, observable states are particles. So symmetry groups and their representations are a basic feature of particle physics.

Historically, the importance of symmetry in particle physics traces back to Hermann Weyl's attempt to unify the forces of electromagnetism and gravity. He suggested that the appropriate symmetries should be changes of spatial scale, or 'gauge'. That approach didn't work out, but Shinichiro Tomonaga, Julian Schwinger, Richard Feynman, and Freeman Dyson modified it to obtain the first relativistic quantum field theory of electromagnetism, based on a group of 'gauge symmetries' $U(1)$. This theory is called quantum electrodynamics.

The next major step was the discovery of the 'eightfold way', which unified eight of the particles that were then considered to be elementary: neutron, proton, lambda, three different sigma particles, and two xi particles. Figure 49 shows the mass, charge, hypercharge, and isospin of each of these particles. (It doesn't matter what these words mean: they are numbers that characterize certain quantum properties.) The eight particles divide naturally into four families, in each of which the hypercharge and isospin are the same, and the masses are nearly the same. The families are:

singlet: lambda
doublet: neutron, proton
doublet: the two xis
triplet: the three sigmas

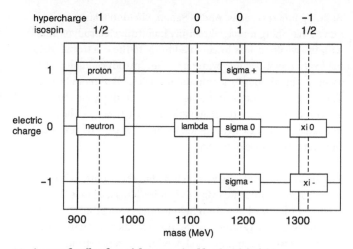

49. A superfamily of particles organized by the eightfold way

where the adjectives indicate how many particles there are in each family.

The eightfold way interpreted this 'superfamily' of eight particles using a particular eight-dimensional irreducible representation of the group **U**(3), whose choice had good physical motivation. Ignoring time breaks the symmetry to a subgroup **SU**(3), which acts on the same eight-dimensional space. This representation of **SU**(3) breaks up into four irreducible subspaces, of dimensions 1, 2, 2, 3. Each of these dimensions corresponds to the number of particles in one of the families. Particles in the same family—that is, corresponding to the same irreducible representation of **SU**(3)—have the same mass, hypercharge, and isospin *because of the **SU**(3) symmetry*. The same ideas applied to a different ten-dimensional representation predicted the existence of a new particle, not known at the time, called the Omega-minus. When this was observed in particle accelerator experiments, the symmetry approach became widely accepted.

Building on these ideas, Abdus Salam, Sheldon Glashow, and Steven Weinberg managed to unify quantum electrodynamics with the weak nuclear force. In addition to the electromagnetic field with its **U**(1) gauge symmetry, they introduced fields associated with four fundamental particles, all of them bosons. The gauge symmetries of this new field form the group **SU**(2), and the combined symmetry group is **U**(1)×**SU**(2), where the × indicates that the two groups act independently. The result is called the electroweak theory.

The strong nuclear force was included in the picture with the invention of quantum chromodynamics. This assumes the existence of a third quantum field for the strong force, with gauge symmetry **SU**(3). Combining the three fields and their three groups led to the standard model, with symmetry group **U**(1)×**SU**(2)×**SU**(3). The **U**(1) symmetry is exact, but the other two are approximate. It is thought that they become exact at very high energies. All three groups contain continuous families of symmetries, and Noether's Theorem tells us that conserved quantities are associated with these. They turn out to be various 'quantum numbers' associated with quarks, such as charge, isospin, and hypercharge. Virtually the whole of particle physics can be accounted for using these basic symmetries of the quantum world.

* * *

One force is still missing: gravity. There ought to be a particle associated with the gravitational field. If it exists, it has been dubbed the graviton. However, unifying gravity with quantum chromodynamics is not just a matter of adding yet another group to the mix. The current theory of gravity is general relativity, and that doesn't fit very neatly into the formalism. Even so, symmetry principles underpin one of the best-known attempts at unification: the theory of superstrings, often called string theory. The 'super' refers to a conjectured type of symmetry known as supersymmetry, which associates to each ordinary particle a supersymmetric partner.

String theories replace point particles by vibrating 'strings', which originally were viewed as circles, but are now thought to be higher-dimensional. Incorporating supersymmetry leads to superstrings. By 1990 theoretical work had led to five possible types of superstring theory, designated types I, IIA, IIB, HO, and HE. The corresponding symmetry groups, known as gauge groups because of the way they act on quantum fields, are respectively the special orthogonal group **SO**(32), the unitary group **U**(1), the trivial group, **SO**(32) again, and $E_8 \times E_8$, two distinct copies of the exceptional Lie group E_8 acting in two different ways. In all five types, space is required to have six extra dimensions—possibly curled up so tightly that we do not observe them, or possibly inaccessible to us because we are confined to a four-dimensional space-time 'brane'.

Soon after, Edward Witten unified all five theories into a single *M*-theory with seven extra dimensions of space. Despite much effort by theorists, no solid evidence for superstring theory has yet been found. Even so, superstrings have greatly enriched mathematics, whatever their fate in fundamental physics will prove to be.

Many alternatives to string theory were and still are being investigated. A new way to calculate properties of collisions between fundamental particles has greatly simplified applications of string theory. It has also revived an earlier attempt to unify gravity with the other three forces of Nature, called supergravity. This was largely abandoned in the 1980s because it was believed to lead to nonsensical infinite quantities, but the new method shows that at least some of the old calculations were misleading. The new approach is called the unitarity method.

The method it is starting to replace has been used for many years. It is based on Feynman diagrams, which represent particle collisions. Even the simplest collision involves infinitely many

Feynman diagrams, because quantum mechanics allows extra particles to appear temporarily and then disappear again. These 'virtual particles' cannot be observed directly, but they create extra terms in the calculation, which is a sum over all possible Feynman diagrams. Richard Feynman introduced this method for quantum electrodynamics, and here the infinite series converge fairly rapidly, so terms that correspond to complicated diagrams can be ignored. But in quantum chromodynamics, the stronger coupling force makes the series converge more slowly, and the number of terms required explodes.

Mysteriously, however, the end result is often very simple. Huge blocks of terms seem to cancel out. Unitarity gets rid of all these terms in one go, by exploiting a basic property of probability: the total probability of all alternatives must be 1. Sums involving millions of Feynman diagrams are replaced by formulas that fit on a single page. The method also uses symmetry principles and new combinatorial ideas.

Supergravity, unlike string theory, represents particles as points. Around 1995 Stephen Hawking suggested taking another look at supergravity, because the calculations that seemed to produce unwanted infinities involved some unsupported assumptions. But it was impossible to perform the calculations more accurately using Feynman diagrams. A collision that included three virtual gravitons, for example, would require summing 10^{20} terms. In 2007 it turned out that the unitary method reduces this number to less than 20. Accurate calculations then became feasible, and it transpired that some of the infinities suspected in the 1980s do not occur. Intriguingly, in the examples studied so far, the graviton behaves like two copies of a gluon. Metaphorically, gravity is like the square of the strong nuclear force. If this feature is true in general, rather than being some coincidence of simple interactions, it will imply that gravity is much more like the other three forces than has been thought. It might even lead to a new, and computationally tractable, unified field theory.

Chapter 8
Atoms of symmetry

One of the greatest scientific achievements of the 19th century was Mendeleev's discovery of the periodic table, which organizes the basic building blocks of matter into sets of substances with similar properties. These building blocks are chemical molecules that cannot be broken up into smaller molecules: in short, atoms. Collectively, they are called elements. By the 20th century it turned out that atoms are themselves composed of smaller subatomic particles, but before that atoms were defined as indivisible particles of matter. The name, in fact, is Greek for 'indivisible': a = not, *temno* = cut. To date 118 elements have been identified, of which ninety-eight occur naturally. The rest have been synthesized in nuclear reactions. All of the latter are radioactive (as are eighteen of the former) and most are very short-lived.

In a loose analogy, every finite symmetry group can be broken up, in a well-defined manner, into 'indivisible' symmetry groups—atoms of symmetry, so to speak. These basic building blocks for finite groups are known as *simple groups*—not because anything about them is easy, but in the sense of 'not made up from several parts'. Just as atoms can be combined to build molecules, so these simple groups can be combined to build all finite groups.

One of the greatest mathematical achievements of the 20th century was the discovery that—to continue the analogy—there is a kind of periodic table for symmetries. This table contains infinitely many groups, but most of them are arranged into families. In addition, a few simple groups do not fit into any family. They are mathematical orphans, destined to live lonely lives; eccentric 'one-offs' known as sporadic groups. There are twenty-six of them.

The proof of the Classification Theorem for Finite Simple Groups, when first completed, ran to about 10,000 pages in mathematical journals. It has since been reworked, exploiting insights acquired along the way to streamline the proof, and it is now estimated that when complete it will be 5,000 pages long. An even more streamlined third-generation proof is under investigation. However, it was always obvious that any proof would have to be unusually lengthy, because the answer itself is complicated. The surprise, if anything, is that it can be done at all; even more so that a mere 5,000 pages are needed.

* * *

In Chapter 4 we described two distinct ways to extract smaller groups from bigger ones, both discovered by the early pioneers of group theory. The most obvious of these concepts is that of a subgroup, which is a subset of a group that forms a group in its own right. The second concept is that of a quotient group, which we saw is associated with a special kind of subgroup, known as a normal subgroup. Recall that an intuitive way to visualize a quotient group is conceptually to colour the group elements. If it is true that whenever elements of two given colours are combined, the colour of the result is always the same, then the colours themselves form a group, and this is the quotient. The corresponding normal subgroup consists of all elements with the same colour as the identity. Every group with more than one element has at least two quotient groups. In one, we colour all elements using the same colour, and the quotient group has only one element. In the other, we colour all elements using distinct colours, and the quotient group is the original group. Neither is terribly interesting. If these are the only quotient groups, then we say that the original group is *simple*.

The smallest simple group, aside from cyclic groups of prime order, is the alternating group $\mathbf{A}_5$ with sixty elements. This group is isomorphic to the group of rotational symmetries of the dodecahedron, discussed in Chapter 3. Table 4 in that chapter contains the basic information for a short, snappy proof that $\mathbf{A}_5$ is simple. The key idea is that if a normal subgroup of some group contains an element, say h, then it must also contain all conjugates $g^{-1}hg$ as g ranges through the entire group. Recall that conjugacy, geometrically speaking, means 'do the same thing at another location'. So symmetries of the same type are conjugate to each other. The sizes of these 'conjugacy classes' are 1, 12, 12, 15, and 20. Any normal subgroup must be a union of some of these classes. Moreover, it must contain the identity (the class with one element), and by Lagrange's Theorem its order must divide 60. So we are seeking solutions of the equation

1 + (a selection from 12, 12, 15, 20) divides 60

and it is easy to show that the only solutions are:

1 = 1
1 + 12 + 12 + 15 + 20 = 60

Therefore the only normal subgroups are the identity and the whole group, and that implies it is simple.

It is possible to apply the same argument directly to $\mathbf{A}_5$ by relating its conjugacy classes to the decomposition of permutations into cycles, and there are other ways to prove that it is simple as well. Modern treatments of Galois Theory use the simplicity of $\mathbf{A}_5$ to prove that quintic equations cannot be solved using radicals. Leaving out a heap of important technicalities, the main idea is that extracting a radical is equivalent to forming a cyclic quotient group of the symmetry group of the equation. If there are no nontrivial proper quotients, there are no cyclic ones, hence no radicals that simplify the equation.

Simple groups are roughly analogous to prime numbers. In number theory, every integer can be written as a product of prime factors; moreover, those factors are unique except for the order in which they appear. There is an analogous statement for finite groups, known as the Jordan–Hölder theorem. It states that any finite group can be broken up into a finite list of simple groups, and that these 'composition factors' are unique except for the order in which they appear. More precisely, for any finite G there exists a chain of subgroups

$$1 = G_0 \subseteq G_1 \subseteq G_2 \subseteq \ldots \subseteq G_r = G$$

each normal in the next, such that every quotient group G_{m+1}/G_m is simple.

For example, if $G = \mathbf{S}_n$ and $n \geq 5$ then such a chain is:

$$1 \subseteq \mathbf{A}_n \subseteq \mathbf{S}_n$$

The composition factors are:

$$\mathbf{A}_n / 1 \cong \mathbf{A}_n \quad \mathbf{S}_n / \mathbf{A}_n \cong \mathbf{Z}_2$$

Properties of $\mathbf{S}_2$, $\mathbf{S}_3$, and $\mathbf{S}_4$ can be used to deduce the solutions of quadratic, cubic, and quartic equations by radicals that were known in Babylon and in Renaissance Italy. Using similar methods in an era before groups were available, Gauss found a ruler-and-compass construction for the regular seventeen-sided polygon. We now interpret his method in terms of composition factors of the multiplicative group of nonzero elements of **GF** (17).

We can recover any number uniquely from its prime factors by multiplying them all together. This is *not* the case for groups. Many different groups can have the same composition factors. So the analogy with prime factorization is very loose. Nevertheless, simple groups play the same prominent role in group theory that primes do in number theory.

A closer analogy is one we have already hinted at: molecules and atoms. Every molecule is composed of a unique set of atoms, but a given set of atoms may correspond to many different molecules. A simple example is ethanol and dimethyl ether. Both are composed of six hydrogen atoms, two carbon atoms, and one oxygen atom. However, those atoms are joined together in two different ways; see Figure 50. This is one justification of the metaphor in which simple groups are the atoms of finite groups.

The search for simple groups occupied the attention of algebraists for over 150 years, starting from the time of Galois. The most obvious such groups are the cyclic groups $\mathbf{Z}_p$ of prime order p. These could not be explicitly recognized as simple groups until 'group' and 'simple' were defined, but the reason why they are simple—primes have no proper factors—goes back to Euclid. Unlike all other simple groups, the cyclic groups are abelian.

Galois found the first nonabelian simple groups in 1832: the two-dimensional projective special linear groups $\mathbf{PSL}_2(p)$,

50. *Left*: Ethanol. *Right*: Dimethyl ether

associated with geometries over finite fields with a prime number $p \geq 5$ of elements. These groups are analogous to the Lie groups $\mathbf{PSL}_2(\mathbf{R})$ and $\mathbf{PSL}_2(\mathbf{C})$, which are groups of 2×2 matrices over the fields $\mathbf{R}$ and $\mathbf{C}$ modulo scalar multiples of the identity, except that $\mathbf{R}$ and $\mathbf{C}$ are replaced by a finite field $\mathbf{GF}(p)$. It was soon recognized that the alternating groups $\mathbf{A}_n$ are simple for $n \geq 5$. The smallest nonabelian simple group is $\mathbf{A}_5$, with order 60. The next smallest is $\mathbf{PSL}_2(\mathbf{GF}(7))$ with order 168.

The next simple groups to be discovered did not fit into any nice family of groups with closely related properties. They were what we now call sporadic groups. In 1861 Émile Mathieu found the first of these, $\mathbf{M}_{11}$ and $\mathbf{M}_{12}$, now named after him. They contain 7,920 and 95,040 elements, respectively. One way to construct them is to employ combinatorial structures known as Steiner systems. For example, a (5, 6, 12) Steiner system is a collection of six-element subsets of a set with twelve elements, having the property that every five-element subset occurs in exactly one of the six-element subsets. There is exactly one such system, up to isomorphism. One way to construct it is to start with $\mathbf{GF}(11)$, the integers modulo 11. This is a finite field since 11 is prime. Add a twelfth point at infinity, ∞. These twelve points form a finite geometry called a *projective line*. There are natural maps from the projective line to itself: the fractional linear transformations (just like Möbius transformations of $\mathbf{C}$):

$$z \ a \ \frac{az+b}{cz+d}$$

where $a, b, c, d \in \mathbf{GF}(11)$ and we interpret $1/0$ as ∞.

To form the six-element subsets, take the set of all squares {0, 1, 3, 4, 5, 9} and apply all possible fractional linear transformations. We obtain a list of 132 subsets, each with six elements. Using the algebra of $\mathbf{GF}(11)$ it is possible to show that each five-element subset occurs in exactly one of these six-element subsets.

The Mathieu group M_{12} can then be defined as the symmetry group of this Steiner system; that is, the group of permutations of $\{0, 1, 2, 3, 4, 5, 6, 7, 8, 9, 10, 11, \infty\}$ that map each six-element subset in the list to another in the list. M_{11} is the subgroup fixing one point. Mathieu also found three other sporadic simple groups in a similar manner. M_{24} is the symmetry group of a (5, 8, 24) Steiner system, M_{23} is the subgroup fixing one point, and M_{22} is the subgroup fixing two points.

The Mathieu groups have relatively large orders, too large for pencil-and-paper listing. However, by the standards of sporadic simple groups the Mathieu groups are tiny. The monster, predicted in 1973 by Bernd Fischer and Robert Griess, and constructed in 1982 by Griess, has

808 017 424 794 512 875 886 459 904 961 710 757 005 754
368 000 000 000

elements—roughly 8×10^{53}. It is the group of symmetries of a curious algebraic structure, the Griess algebra.

Despite these complexities, the early discoveries were representative of the complete list. We now know that every finite simple group is one of the following:

- cyclic groups of prime order.
- alternating groups $\mathbf{A}_n$ for $n \geq 5$.
- sixteen families of groups, analogues of simple Lie groups that replace $\mathbf{R}$ or $\mathbf{C}$ by finite fields, called groups of Chevalley type after Claude Chevalley. Many of these families had been constructed previously, but Chevalley found a unified description that led to new families. Nine of these families are now called Chevalley groups. Defining them is not just a matter of taking matrix groups and changing the field, but that's what motivated the idea.
- twenty-six sporadic groups—one-offs like the Mathieu groups.

Table 8. The twenty-six sporadic finite simple groups

Symbol	Name	Order
M_{11}	Mathieu group	$2^4.3^2.5.11$
M_{12}	Mathieu group	$2^6.3^3.5.11$
M_{22}	Mathieu group	$2^7.3^2.5.7.11$
M_{23}	Mathieu group	$2^7.3^2.5.7.11.23$
M_{24}	Mathieu group	$2^{10}.3^3.5.7.11.23$
J_1	Janko group	$2^3.3.5.7.11.19$
HJ	Hall–Janko group	$2^7.3^3.5^2.7$
HJM	Higman–Janko–McKay group	$2^7.3^5.5.17.19$
J_4	Janko group	$2^{21}.3^3.5.7.11^3.23.29.31.37.43$
Co_1	Conway group	$2^{21}.3^9.5^4.7^2.11.13.23$
Co_2	Conway group	$2^{18}.3^6.5^3.7.11.23$
Co_3	Conway group	$2^{10}.3^7.5^3.7.11.23$
Fi_{22}	Fischer group	$2^{17}.3^9.5^2.7.11.13$
Fi_{23}	Fischer group	$2^{18}.3^{13}.5^2.7.11.13.17.23$
Fi_{24}	Fischer group	$2^{21}.3^{16}.5^2.7^3.11.13.17.23.29$
HS	Higman–Sims group	$2^9.3^2.5^3\ 7.11$
McL	McLaughlin group	$2^7.3^6.5^3\ 7.11$
He	Held group	$2^{10}.3^3.5^2\ 7^3\ 17$
Ru	Rudvalis group	$2^{14}.3^3.5^3\ 7.13.29$
Suz	Suzuki group	$2^{13}.3^7.5^2\ 7.11.13$
ONS	O'Nan–Sims group	$2^9.3^4.5.7^3.11.19.31$
HN	Harada–Norton group	$2^{14}.3^6.5^6.7.11.19$

LyS	Lyons–Sims group	$2^8.3^7.5^6.7.11.31.37.67$
Th	Thompson group	$2^{15}.3^{10}.5^3.7^2.13.19.31$
B	Baby monster	$2^{41}.3^{13}.5^6.7^2.11.13.17.19.23.31.47$
M	Monster	$2^{46}.3^{20}.5^9.7^6.11^2.13^3.17.19.23.29.$ $31.41.47.59.71$

A list of the families is not especially informative on its own; details can easily be found on the Internet, for example in Wikipedia. Table 8 lists the sporadic groups, and shows why 'sporadic' is a sensible name. All but the final two groups are named after whoever discovered them.

This classification was obtained between 1955 and 2004 through the joint efforts of about a hundred mathematicians, ultimately following a programme proposed by Daniel Gorenstein. As already remarked, a more streamlined version has been found and further simplifications are in train. The sheer complexity of the classification and its enormous proof are a testament to the power of mathematics and the dedication and persistence of its practitioners. It is one of the most impressive high points in our growing understanding of symmetry.

* * *

Initially the classification of the finite simple groups was an end in itself. It was obviously important, fundamental information on which future mathematicians could build. What they would build was necessarily unclear; if we knew where research was going to lead, it wouldn't be research. There was some speculation about potential applications, but until the classification was complete, these had to remain speculative. Now that the classification has been obtained, applications are already appearing. They use the classification as a crucial part of the proof of results that do not explicitly refer to simple groups. Some are in areas outside group theory.

In 1983 the classification was used to prove a conjecture stated by Charles Sims in 1967, which provides a bound for the size of certain subgroups of primitive permutation groups. The proof required detailed information about the groups of Chevalley type, but did not involve the sporadic groups.

Another application is a link between simple groups and primes, related to maximal subgroups: subgroups that are smaller than the entire group, and such that no subgroup lies strictly between the two. The index of a subgroup is the order of the group divided by the order of the subgroups. In 1982 Peter Cameron, Peter Neumann, and D. N. Teague used the classification to prove that the number of integers up to a given size x, that are the indices of maximal subgroups (other than those of index n in $\mathbf{S}_n$ and $\mathbf{A}_n$), is asymptotic to $2x/\log x$, which means that the ratio of the exact number to the formula tends to 1 as x tends to infinity.

A third application, whose importance in computer science is growing rapidly because of links to error-correcting codes, is the topic of expander graphs. A graph is a collection of vertices joined by edges. A graph is an expander if the number of vertices that are adjacent to a given subset of vertices, but not in that subset, is at least the size of the subset multiplied by some fixed nonzero constant. The subset should not be too big—at most half the number in the entire graph—because otherwise the number of adjacent vertices becomes too small. An *expander family* is a series of expander graphs, whose sizes tend to infinity, which all have the same constant.

It has been known for some time that many expander families exist, but the proof was based on the probability of a random graph being an expander, and did not produce explicit examples. That has now changed, thanks to the classification theorem, which shows that certain graphs associated with finite simple groups are expanders. They are known as Cayley graphs, and they depend on choosing a set of generators for the group: enough elements to obtain all of the others by multiplying them together. In 1989 László Babai, William Kantor, and Alex Lubotzky conjectured that for every positive constant, there exists a number k such that every non-cyclic finite simple group has a set of at most k generators whose Cayley graph is an expander with that constant. This has now been proved by the combined efforts of a number of mathematicians, the first breakthrough being obtained by Martin Kassabov in 2007.

* * *

The story of symmetry, the mathematics that it led to, and the uses to which the resulting theories have been put, show how simple but deep concepts can lead to immensely powerful theories and major scientific advances. However, the path from the original concept to those advances is not a straightforward dash into new territory. It involves tentative forays into the unknown, a lot of 'following your nose' by pursuing ideas that have the right *feel* to them, long before anyone can guarantee their value, and the ability to recognize a good idea. There can be lengthy excursions into abstract generalities, with no obvious immediate use, which eventually prove their worth. Nature, science, and theoretical mathematics can join together to provide new insights into the universe that we inhabit. Above all, the quest to understand symmetry provides a wonderful example of how the beauty of the natural world can lead to beautiful science and beautiful mathematics.

Further reading

Online reading

http://en.wikipedia.org/wiki/Symmetry
http://en.wikipedia.org/wiki/Galois_theory
http://en.wikipedia.org/wiki/Group_theory
http://mathworld.wolfram.com/GroupTheory.html
http://en.wikipedia.org/wiki/Wallpaper_group
http://xahlee.org/Wallpaper_dir/c5_17WallpaperGroups.html
http://www.patterninislamicart.com/
http://sillydragon.com/muybridge/Plate_0675.html
http://www.flickr.com/photos/boston_public_library/
    collections/72157623334568494/
http://www.apple.com/science/insidetheimage/bzreaction/images.html
http://en.wikipedia.org/wiki/Simple_Lie_group
http://en.wikipedia.org/wiki/List_of_simple_Lie_groups
http://en.wikipedia.org/wiki/Classification_of_finite_simple_groups
http://en.wikipedia.org/wiki/List_of_finite_simple_groups

Offline reading

Syed Jan Abas and Amer Shaker Salman. *Symmetries of Islamic Geometrical Patterns*, World Scientific, Singapore 1995.

R. P. Burn. *Groups: A Path to Geometry*, Cambridge University Press, Cambridge 1985.

John H. Conway, Hedie Burgiel, and Chaim Goodman-Strauss. *The Symmetries of Things*, A. K. Peters, Wellesley, MA 2008.

Keith Critchlow. *Islamic Patterns*. Thames and Hudson, London 1976.

Harold M. Edwards. *Galois Theory*, Springer, New York 1984.

D. J. H. Garling. *A Course in Galois Theory*, Cambridge University Press, Cambridge 1986.

Martin Golubitsky and Michael Field. *Symmetry in Chaos* (2nd edn), SIAM, Philadelphia, PA 2009.

Martin Golubitsky and Ian Stewart. *The Symmetry Perspective*, Progress in Mathematics **200**, Birkhäuser, Basel 2002.

Istaván Hargittai and Magdolna Hargittai. *Symmetry: A Unifying Concept*, Shelter Publications, Bolinas, CA 1994.

John F. Humphreys. *A Course in Group Theory*, Oxford University Press, Oxford 1996.

E. H. Lockwood and R. H. Macmillan. *Geometric Symmetry*, Cambridge University Press, Cambridge 1978.

Henry McKean and Victor Moll. *Elliptic Curves*, Cambridge University Press, Cambridge 1997.

Hans Meinhardt. *The Algorithmic Beauty of Sea Shells*, Springer, Berlin 1995.

Peter M. Neumann, Gabrielle A. Stoy, and Edward C. Thompson. *Groups and Geometry*, Oxford University Press, Oxford 1994.

Ernö Rubik, Tamás Varga, Gerazon Kéri, György Marx, and Tamás Vekerdy. *Rubik's Cubic Compendium*, Oxford University Press, Oxford 1987.

David Singmaster. *Notes on Rubik's Magic Cube*, Penguin Books, Harmondsworth 1981.

Ian Stewart. *Galois Theory* (3rd edn), CRC Press, Boca Raton, FL 2003.

Ian Stewart. *Why Beauty is Truth*, Basic Books, New York 2007.

Ian Stewart and Martin Golubitsky. *Fearful Symmetry: Is God a Geometer?*, Blackwell, Oxford 1992. Reprinted Dover Publications, Mineola, NY 2011.

Thomas M. Thompson. *From Error-Correcting Codes Through Sphere Packings to Finite Simple Groups*, Mathematical Association of America, Washington DC 1983.

Jean-Pierre Tignol. *Galois' Theory of Algebraic Equations*, Longman, Harlow 1987.

Hermann Weyl. *Symmetry*, Princeton University Press, Princeton, NJ 1952.

Index

Index

Symmetry

SOCIAL MEDIA
Very Short Introduction

Join our community

www.oup.com/vsi

- Join us online at the official Very Short Introductions **Facebook** page.
- Access the thoughts and musings of our authors with our online **blog**.
- Sign up for our monthly **e-newsletter** to receive information on all new titles publishing that month.
- Browse the full range of Very Short Introductions online.
- Read **extracts** from the Introductions for free.
- Visit our library of **Reading Guides**. These guides, written by our expert authors will help you to question again, why you think what you think.
- If you are a teacher or lecturer you can order inspection copies quickly and simply via our website.

Visit the Very Short Introductions website to access all this and more for free.
www.oup.com/vsi

ONLINE CATALOGUE
A Very Short Introduction

Our online catalogue is designed to make it easy to find your ideal Very Short Introduction. View the entire collection by subject area, watch author videos, read sample chapters, and download reading guides.

http://fds.oup.com/www.oup.co.uk/general/vsi/index.html